I0014144

Rajani Shankar Sadasivam

An Architecture Framework for Process-Personalized Composite Services

Rajani Shankar Sadasivam

An Architecture Framework for Process-Personalized Composite Services

Service-oriented Architecture, Web Services, Business-Process Engineering, and Human Interaction Management

VDM Verlag Dr. Müller

Impressum/Imprint (nur für Deutschland/ only for Germany)

Bibliografische Information der Deutschen Nationalbibliothek: Die Deutsche Nationalbibliothek verzeichnet diese Publikation in der Deutschen Nationalbibliografie; detaillierte bibliografische Daten sind im Internet über http://dnb.d-nb.de abrufbar.

Alle in diesem Buch genannten Marken und Produktnamen unterliegen warenzeichen-, marken- oder patentrechtlichem Schutz bzw. sind Warenzeichen oder eingetragene Warenzeichen der jeweiligen Inhaber. Die Wiedergabe von Marken, Produktnamen, Gebrauchsnamen, Handelsnamen, Warenbezeichnungen u.s.w. in diesem Werk berechtigt auch ohne besondere Kennzeichnung nicht zu der Annahme, dass solche Namen im Sinne der Warenzeichen- und Markenschutzgesetzgebung als frei zu betrachten wären und daher von jedermann benutzt werden dürften.

Coverbild: www.purestockx.com

Verlag: VDM Verlag Dr. Müller Aktiengesellschaft & Co. KG
Dudweiler Landstr. 125 a, 66123 Saarbrücken, Deutschland
Telefon +49 681 9100-698, Telefax +49 681 9100-988, Email: info@vdm-verlag.de
Zugl.: Birmingham, Alabama, University of Alabama at Birmingham, Diss,, 2007

Herstellung in Deutschland:
Schaltungsdienst Lange o.H.G., Zehrensdorfer Str. 11, D-12277 Berlin
Books on Demand GmbH, Gutenbergring 53, D-22848 Norderstedt
Reha GmbH, Dudweiler Landstr. 99, D- 66123 Saarbrücken
ISBN: 978-3-639-08724-6

Imprint (only for USA, GB)

Bibliographic information published by the Deutsche Nationalbibliothek: The Deutsche Nationalbibliothek lists this publication in the Deutsche Nationalbibliografie; detailed bibliographic data are available in the Internet at http://dnb.d-nb.de.

Any brand names and product names mentioned in this book are subject to trademark, brand or patent protection and are trademarks or registered trademarks of their respective holders. The use of brand names, product names, common names, trade names, product descriptions etc. even without
a particular marking in this works is in no way to be construed to mean that such names may be regarded as unrestricted in respect of trademark and brand protection legislation and could thus be used by anyone.

Cover image: www.purestockx.com

Publisher:
VDM Verlag Dr. Müller Aktiengesellschaft & Co. KG
Dudweiler Landstr. 125 a, 66123 Saarbrücken, Germany
Phone +49 681 9100-698, Fax +49 681 9100-988, Email: info@vdm-verlag.de

Copyright © 2008 VDM Verlag Dr. Müller Aktiengesellschaft & Co. KG and licensors
All rights reserved. Saarbrücken 2008

Produced in USA and UK by:
Lightning Source Inc., 1246 Heil Quaker Blvd., La Vergne, TN 37086, USA
Lightning Source UK Ltd., Chapter House, Pitfield, Kiln Farm, Milton Keynes, MK11 3LW, GB
BookSurge, 7290 B. Investment Drive, North Charleston, SC 29418, USA
ISBN: 978-3-639-08724-6

DEDICATION

I dedicate this book to my family.

ACKNOWLEDGMENTS

First, I thank my advisor, Dr. Murat M. Tanik from the Electrical and Computer Engineering (ECE) Department at the University of Alabama at Birmingham (UAB), for his invaluable advice, kind assistance, and contribution towards my dissertation. Without him, this dissertation would not be possible. Dr. Murat M. Tanik has provided ample support throughout my years at the University of Alabama at Birmingham (UAB). He has been instrumental in my growth by providing me with opportunities to participate and contribute in several initiatives, such as proposal writing, research projects, and business opportunities.

I thank Dr. Murat N. Tanju, Accounting and Information Sciences at UAB, for his critical evaluation of my work. He patiently provided feedback for improving my qualifying and final defense presentation. Dr. Tanju has also supported my activities through my stay at UAB.

My sincere gratitude is extended to my committee members, Drs. Gregg L. Vaughn and Gary J. Grimes from the ECE Department, Dr. Barrett R. Bryant from the Computer and Information Sciences Department, and Dr. Jeffrey H. Kulick from the ECE Department at the University of Alabama in Huntsville (UAH). I thank them for their kind guidance and suggestions on improving the dissertation, especially by providing important recommendations on project scope, thesis rigor, and presentation clarity.

I thank Dr. Chittoor V. Ramamoorthy, Professor Emeritus at the Electrical Engineering and Computer Sciences Department at UC Berkeley, for serving on my committee. Dr. Ramamoorthy's seminal work on service enterprise forms the basis for my dissertation. Despite his stature in the field of computer sciences, Dr. Ramamoorthy has always amazed me with his humility and open mindedness. Dr. Ramamoorthy has always supported my work during our meetings.

I thank Professor Leon Jololian, Computer Science Department at the New Jersey City University, for his advice, which represented an invaluable contribution to and support for the dissertation. I would like to thank my colleagues, the component-based software engineering research group members, Bunyamin Ozaydin, Dr. Ozgur Aktunc, and Dr. Urcun J. Tanik.

I thank my employers for their support during my years at UAB. Dr. Jill Gemmill and John-Paul Robinson supported my work during my years (2002 to 2004) at the department of Academic Computing. I thank Ramesh Puljala, Aditya Srinivasan, Pravin Joshi, and Silbia Peechakara, my colleagues at the department of Academic Computing, for their support.

I thank my employers, Dr. Jeroan J. Allison, Dr. Linda Casebeer, and Robert Kristofco, for their support of my work during my years (2004 to the present) at the Division of Continuing Medical Education. I especially thank Katie Crenshaw for accommodating my frantic schedule during my dissertation writing. Without her support, I would not have been able to complete my dissertation at this time. I thank my colleagues at CME, Raju Datla, Ramkumar Santhanakrishnan, Periyakaruppan Krishnamoorthy, and Pam Clark for their support.

I thank Dr. David B. Allison, Section of Statistical Genetics at UAB, for providing an opportunity to work with him on the statistical wrapper and other statistical genetics projects at UAB. I thank Dr. Donna K. Arnett and Dr. Laura K. Vaughan, Department of Epidemiology, for providing an opportunity to work on the gene linkage identification project at UAB. Dr. Arnett's project is used as a case study in my dissertation. I thank Dr. John L. Hartman IV, Department of Genetics at UAB, for providing me an opportunity to work on the research integration of his lab processes. Dr. Allison, Dr. Arnett, Dr. Vaughan, and Dr. Hartman's support was instrumental in arranging the collaboration with Oracle's technology and Microsoft's bioinformatics division at UAB.

I thank Dr. Jeffrey G. Gray, Computer and Information Sciences Department at UAB, for providing the opportunity to present my work in the IEEE Alabama computer seminar series.

I thank Dr. Thomas C. Jannett, the graduate program director of the Electrical and Computer Engineering Department at UAB, for streamlining the process of completion of my dissertation. I also thank him for providing me the opportunity to present my work in the EE601 seminar class. I thank Maria Whitmire and Sandra Muhammad for their support in paperwork and other administrative needs.

Finally, I thank my family for their patience, support, and understanding in every step of my life. I especially thank my wife Gayathri Sundar for her support and understanding during the completion of my dissertation.

TABLE OF CONTENTS

Page

vii

LIST OF TABLES

LIST OF FIGURES

LIST OF ABBREVIATIONS

AI	Artificial Intelligence
AOSD	Aspect Oriented Software Development
BPD	Business Process Diagram
BPEL	Business Process Execution Language
BPEL4WS	Business Process Execution Language for Web Services
BPM	Business Process Management
BPMI	Business Process Modeling Initiative
BPML	Business Process Modeling Language
BPMN	Business Process Management Notation
BPMS	Business Process Management Systems
BPQL	Business Process Query Language
BPSM	Business Process Semantic Markup
CIM	Computation Independent Model
CIMOSA	Computer Integrated Manufacturing Open Systems Architecture
CLIPS	C Language Integrated Production System (CLIPS)
CPP	Composite Process-Personalization
CoMoMAS	Conceptual Modeling of Multi Agent Systems
COS	Communication Semantics
DAML	DARPA Agent Markup Language
DAML-S	DARPA Agent Markup Language – Services
DARPA	Defense Advanced Research Projects Agency
EEAF	Extended Enterprise Architecture Framework
EM	Enterprise Model
FEAF	Federal Enterprise Architecture Framework
GLIP	Gene Linkage Identification Process
HTN	Hierarchical Task Network
ICARIS	Infrastructure for Composability at Runtime of Internet Services
INS	Infrastructure Semantics
KS	Knowledge Semantics

MDA	Model Driven Architecture
MIME	Multipurpose Internet Mail Extensions
NCBI	National Center for Biotechnology Information
NSF	National Science Foundation
OIL	Ontology Inference Layer
OMG	Object Management Group
OMIM	Online Mendelian Inheritance in Man
OWL-S	Semantic Markup for Web Services
P^2_{DM}	Process-Personalized Development Model
P^2_{EMS}	Process-Personalized Enterprise Management System
$P^2_{FRAMEWORK}$	Process-Personalized Composite-services Framework
PIM	Platform Independent Model
PSM	Platform Specific Model
QoS	Quality of Service
RDF	Resource Description Framework
REST	Representational State Transfer
RIM	Ramamoorthy's Interaction Model
ROS	Roles Semantics
RPC	Remote Procedure Call
RPC	Remote Procedure Call
RUS	Rules Semantics
SAS	Statistical Analyses Software
SHOP	Simple Hierarchical Ordered Planner
SOA	Service-oriented Architecture
SODA	Societies in Open and Distributed Agent spaces
TAS	Tools-As-Services
TOGAF	The Open Group Architecture Framework
UDDI	Universal Description, Discovery and Integration
UML	Unified Modeling Language
UPS	Users-Profile Semantics
V&V	Verification and Validation
WAN	Wide Area Network
WCS	Weather Composite Service
WfMC	Workflow Management Coalition

WS-CDL	Choreography Description Language
WSDL	Web Services Description Language
WSFL	Web Services Flow Language
WWF	Windows Workflow Foundation
WWW	World Wide Web
XML	Extensible Markup Language
XPDL	XML Process Definition Language
xUML	Executable UML

I. INTRODUCTION

This book addresses the integration challenges of large systems development by developing a composition, integration, and personalization framework. A large system can be defined as a system that is composed of interrelated subsystems [1]. Composite services development can be defined as a large systems development approach in which the focus is on specifying, discovering, selecting, and integrating services. A service in this context refers to the resources that perform work to satisfy the needs of the large system [2]. Development of large systems with a composite service approach is important because we can leverage the rich set of composite services tools, technologies, and approaches [3], [4]. The rich set of composite services tools, technologies, and approaches come from the industrial and academic track, which are complementary to each other [3]. For instance, the industrial track of composite services research has provided several sophisticated tools and standards for developing composite services. The academic track has developed agent-based approaches for automated discovery and composition. However, despite these advances, the integration and development of large systems remain problematic [5]-[8]. The core contribution of this book is as follows:

- First, a criticial reason for the integration problem is identified as the lack of composite, integrated, and personalized support for human interaction with processes, the composite process-personalization (CPP) challenge of large systems development.

- Second, the composite $P^2_{FRAMEWORK}$ is developed, which provides systematic guidance for the development of composite, integrated, and personalized large systems. The composite $P^2_{FRAMEWORK}$ realizes large systems integration with CPP based on the following: The CPP dimensions, which identify the semantic and syntactic aspects of CPP, and the service-agent model, an abstraction that provides a modeling approach for the development of large systems with CPP. The service-agent model supports both the semantic and syntactic dimensions of CPP. A concept map is developed that visualizes and articulates the relationship of the CPP semantics with the service-agents. The relationships articulated by the concept map can be used to model and develop large systems with CPP.

- Third, a process mining approach is developed that can be used for process analysis and optimization in combination with the CPP semantics. The service-agent model can be leveraged for the necessary semantics.

The composite $P^2_{\text{FRAMEWORK}}$ primarily builds on process-oriented development of composite services for large systems integration with CPP [9]-[11]. Process-oriented development of composite services is an approach emphasizing the use of processes throughout the composite services life cycle. This book extends my work on large systems integration, discussed in [12]-[23], and applied to an assortment of large systems, such as sensor systems, agent systems, life sciences research integration systems, digital library systems, and enterprises. A part of this book's work was also applied to developing a CPP patent [24]. Two case studies are used to demonstrate and validate the guidance offered by the composite $P^2_{\text{FRAMEWORK}}$.

The type of large systems discussed in this book is enterprises. Enterprises are emblematic of large systems whose relationships are between subsystems of four types. The four types of subsystems are software, hardware, netware, and peopleware [25]. Enterprise development applies to the development of service enterprises, virtual organizations, cyberinfrastructure, and grid systems [2], [17], [23], [26], [27].

The breakdown of this book is as follows: In Chapter I, the lack of CPP support is identified as the integration challenge that this book addresses. The motivation and approach of this book is also described.

In Chapter II and III, a discussion of the concepts used in the development of the composite $P^2_{\text{FRAMEWORK}}$ is provided. First, in Chapter II, a definition of composite services is provided. The life cycle of composite services development is introduced. The different architectures for composite services development are discussed. An example of process-oriented composite services development with service-oriented architecture (SOA) and Web services is also provided. Finally, a brief discussion of the agent concept used in the service-agent model is provided.

In Chapter III, an overview of process engineering of composite services is provided. Since the term "process" has several different connotations, a definition is provided to clarify its use in this book. The process life cycle that forms an integral part of the composite P²$_{\text{FRAMEWORK}}$ is described. The different process technologies and process formalisms that can be used in process engineering are described. The process-oriented composite services development approaches are classified.

An overview of enterprise architecture frameworks is also provided in Chapter III. The six types of semantics identified in the CPP dimensions and their integration are discussed. The task system model that is used to model the processes of the case study examples is discussed [28]-[33]. The process mining approach for analysis of composite services is discussed.

In Chapter IV, the dimensions required to address the CPP challenge are defined. The composite P²$_{\text{FRAMEWORK}}$, an architecture framework for guiding composite services development with emphasis on CPP is described. The service-agent model is also described. Two case studies that demonstrate and validate the composite P²$_{\text{FRAMEWORK}}$ guidance are provided. A CPP development model is also described. The process mining approach is demonstrated.

In Chapter V, concluding remarks are provided on the book and an overview of the future research that is possible on this topic is provided.

A. Motivation – The Integration Challenge

The motivation for this book is to reduce the process gap that occurs in large systems development. The process gap occurs when the implemented composite service delivers a process that does not meet the needs of the users [34]. In this section, a critical reason for the occurrence of the process gap in large systems development is discussed using enterprises as representative of large systems.

The Internet has caused a paradigm shift in the operation of enterprises [13], [22], [35]-[37]. Consequently, enterprises have had to shift their business model from monolithic, centralized operations to Internet-based, collaborative, and distributed operations. This Internet-based model provides significant opportunities to the enterprise, such as enabling enterprises to establish relationships between each other virtually and dynamically [35], [37], [38]. Consequently, the possibility of bringing together, quickly and efficiently, several organizations that are first-rate within their core capabilities is created, thereby creating a composite enterprise that is first-rate overall [39]. The Internet-based model also has created significant challenges to remain competitive in the Internet market [13], [36], such as the need to reduce time-to-market with new products and services, provide better services and sales processes, and offer lower operational, production, and inventory costs.

Therefore, enterprises face a continuous need for integration of enterprise processes [5]-[8]. Composite services technologies have evolved to address the integration challenges [9], [40]-[42]. In particular, the advances in services and process technologies have enabled us to develop enterprises by composing services quickly and effectively [21], [43]-[47]. In spite of these technological advances, we have not met the integration challenge [5]-[8]. Noted author Peter Fingar makes the following observation [5], [48]:

> Although the past five years have witnessed great progress
> in the theory and practice of business process management, deployments have so far been mostly tactical. Reviewing the case studies that abound at business process management (BPM) vendors' Web sites, many business processes have been revamped and there are return of investment (ROI) stories to tell. That is good; but typical BPM deployments to date have been limited in scope, applying to improvements in specific business functions. Today, however, what executive wants to settle for small internal gains in Department X or Department Y, when there is a battle for survival raging? The full potential of BPM is about 'enterprise business processes' and 'value-chain business processes' not technical improvements here or there, or streamlining individual functions in the company.

The integration challenge cannot be met until we understand its full scope [17], [34]. Consider Ramamoorthy's interaction model (RIM) of enterprises [2]. RIM describes three types of enterprise interactions [2]: 1) Mechanistic interactions among service processes, 2) interactions between service processes and individuals, and 3) interactions between service processes and teams. RIM implies that the scope of the integration problem is more than just the integration of mechanistic processes. It encompasses the integration of processes with human interactions. However, our technologies are only suitable for integration of mechanistic processes [34]. They are not capable of handling processes with human interaction. I refer to this integration challenge as the CPP challenge [17], [21], [24].

The lack of CPP support in large systems development causes the process gap in enterprises. Fig. 1 depicts that a process gap exists between user needs and the implemented composite services. We must reduce the process gap to deliver composite services that match the user need. The semantics of the user needs come from the problem domain, that is, the semantics describe the user tasks and the resources needed for those tasks. The difficulty in capturing, and manging the semantics of the user needs is a major factor in the occurrence of process gap in large systems development. The difficulties exist because of multiple reasons, such as lack of understanding of the problem domain by the developer, lack of understanding of the scope of the integration problem, use of ad hoc approaches, and the lack of technology support as discussed in [34].

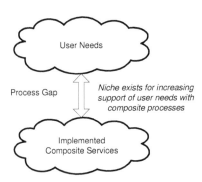

Fig. 1. Process gap of enterprise processes.

Consider the composite process depicted in Fig. 2(a), which is used to illustrate the difficulties in capturing the semantics of user needs. It depicts a process of epidemiology research [21], [49]. For explaining the problem of capturing the semantics of user needs, only a high-level overview of the process is provided. The first stage in the research process is experimentation. The data collection and data analysis stage performed by the assistant and the analyst respectively, follows the experimentation stage. A multitude of tools such as Microsoft Excel, databases, and Statistical

5

Analysis Software (SAS) are used for the purpose of data collection and analysis. The gene linkage identification process (GLIP) follows the data analysis stage. An expanded depiction of the GLIP is provided in Fig. 2(b). The GLIP constitutes the steps to find candidate genes for further studies in other organisms, such as rats and mice. A detailed explanation of GLIP is provided in Chapter IV and also discussed in [21]. The post-doctoral fellows assigned to the researcher perform the gene-linkage identification. Finally, the list of identified candidate genes is forwarded to the collaborators of the researchers for experiments in rats and mice.

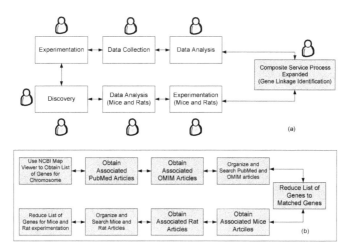

Fig. 2. Epidemiology research process.

The research process includes a set of tasks and resources for successful execution: First, each stage in the research process includes several tasks that must be performed; an example is depicted in Fig. 2b, which expands the GLIP. Second, it involves the use of different tools including Microsoft Excel, databases, SAS, and those that are provided on the National Center for Biotechnology Information (NCBI) Web site, such as the Map Viewer tool, PubMed, and Online Mendelian Inheritance in Man (OMIM) databases [50]. Third, it involves collaborations between the team members for the successful transfer of resources and tasks. Fourth, it involves the knowledge of the different users for successfully performing their tasks. For example, domain knowledge is required for providing the different parameters required by the tools and setting conditions for the identification of candidate genes.

The research process example illustrates the different types of semantics of user needs that must be captured, configured, and managed for addressing large systems integration needs, demonstrating the need for a composite and systematic framework that guides the integration of large systems. The process gap effect can be seen in the integration confusion caused by the divergence of integration technologies used in enterprises. Two groups of integration technologies have been developed to address the integration of enterprise processes. One is the BPM technologies group that address the automation of mechanistic processes [51], [52]. Another is the groupware technologies group, including knowledge management tools, collaborative systems, and content management systems, that address the integration needs of human participants in a process [53]. The divergence of technologies, instead of solving the problem, have caused greater confusion in enterprises [34], [48]. The reason is that these two groups of technologies exist in function silos, which implies that they do not integrate well with each other leading to a lack of integrated information to the user. The ineffective use of collaborative technologies results in collaboration that occurs out of context of enterprise processes, resulting in the inefficient operation of enterprises [34]. Fig. 3 depicts the integration confusion in enterprises.

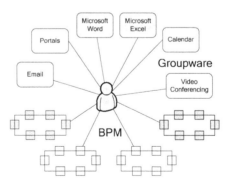

Fig. 3. Integration confusion in enterprises.

B. Composite Process-Personalized Systems and Tools-as-Services

In this section, an example is used to illustrate the needs of CPP. Consider that users in highly interactive working scenarios may prefer certain tools for accessing data and accomplishing tasks. For example, in the composite process depicted in Fig. (2b), the user requires four tools to complete his task. The user's process is depicted in Fig. 4. He uses the NCBI Map Viewer tool [50] to obtain a list of genes for his study, uses Microsoft Excel to view and manipulate the list of genes, uses the PubMed and OMIM Web sites [50] to obtain the articles associated with the list of genes, and Microsoft Word to organize and search through the articles. This implies that the user's infrastructure needs are a combination of the data sources, such as the gene, PubMed, and OMIM databases, and the user's tools, such as the NCBI Map Viewer, Microsoft Word and Microsoft Excel.

In this example, as in most integration efforts, the integration scope is restricted only to the data needs of the user. For example, NCBI has provided a unified way of accessing the genes, PubMed, and OMIM data. However, the user tools are not seamlessly integrated with the data, which leaves the user to transfer the data between the tools. This process can be time-intensive and error-prone, resulting in a poor user experience that implies the lack of CPP.

There are two issues in addressing the user's needs in this scenario. First, as discussed in the motivation, there is a need for an approach to comprehensively capture, configure, and manage the semantics of user needs in composite services development. Second, there is a need for an approach to compose tools-as-services (TAS), which enables the tools to be seamlessly integrated with the data. The first issue is addressed in this dissertation. TAS is addressed in [17], [21] and also in the CPP patent [24].

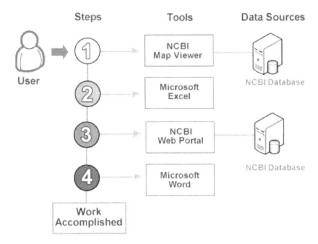

Fig. 4. Example depicting need for Tools-as-Services.

C. Book Approach

The CPP challenge is the lack of support for the integration of human interactions in large systems development. The integration problem occurs because of difficulties in capturing, configuring, and managing the semantics of user needs with current process-oriented composite services approaches. The lack of CPP causes two significant and related problems. First, it causes integration problems in large systems, resulting in the process gap. Second, the process analysis of large systems cannot be comprehensive because we have not incorporated all of the necessary semantics and are operating with an incomplete view of the system.

Unless enterprises can integrate and analyze their processes effectively, they will be inefficient and may be ineffective. In time, their process will become more complex to manage [2], [34], [54]-[58]. A systematic framework that guides the composite, integrated, and personalized integration of large systems is required. Interaction of users with processes and support for their decision-making should be natural extensions of the design and development of systems and not an afterthought or an add-on. This implies that CPP should be an intricate part of the engineering and development of enterprise systems.

This book addresses the lack of CPP support by developing a composition, integration, and personalization framework, the composite $P^2_{FRAMEWORK}$, which provides systematic guidance for the integration of large systems. Under the guidance of the composite $P^2_{FRAMEWORK}$, large systems with CPP are realized based on the CPP dimensions and the service-agent model. The semantic aspects of the CPP dimensions provide guidance for the capture of user needs. The syntactic aspects of the CPP dimensions specify the need for flexible integration and automation of large systems processes, when applicable. The syntactic aspects are addressed in the composite $P^2_{FRAMEWORK}$ by the improved capture and retention of semantics, which provide support for enhanced composition, analysis, and optimization of composite services. A concept map is developed that visualizes and articulates the relationship of the CPP semantics with the service-agents for modeling and development of large systems with CPP. Concept maps are used because they provide a strong foundation for the development of ontology and DSL [59]-[61].

A process-mining-approach based on Conant's decomposition approach [62], [63] is also developed for analysis of large systems that can leverage the service-agent model for the necessary semantics.

Two case studies are used to demonstrate and validate the guidance offered by the composite $P^2_{FRAMEWORK}$. The first case study is a weather composite service (WCS) process with a human interaction task. The second case study is a part of the composite epidemiology research process for identifying candidate genes for obesity research. This case study was developed in collaboration with the Department of Epidemiology at the University of Alabama at Birmingham (UAB), and involves complex interactions between the different participants in the research process. Both case studies demonstrate the guidance provided by the framework for developing large systems with CPP.

D. Conclusion

In this chapter, the lack of CPP support was identified as the integration challenge that this book addresses. The motivation and approach of this book was also described.

II. INTRODUCTION TO COMPOSITE SERVICES

In this chapter, a definition of composite services is provided. The life cycle of composite services development is introduced. The different architectures for composite services development are discussed. An example of process-oriented composite services development with service-oriented architecture (SOA) and Web services is also provided. Finally, a brief discussion of the agent concept used in the service-agent model is provided.

A. *Composite Services*

A composite service is composed of a variety of services to realize enterprise needs. The processes provide the syntactic and semantic elements of composite services to describe the enterprise need. The syntactic elements control, coordinate, and orchestrate the communication of the tasks that realize the composite service. The semantic elements capture the "meaning" of a system, such as representing the work to be accomplished, the person or role that performs the work, and the steps to accomplish the work. The tasks of a process can use multiple resources, such as an automated service, a tool, or a human activity.

Consider the equation $\vec{F} = M \cdot \vec{A}$ as a representation of composite services. $\vec{F}$ is a combination of three vectors, $\vec{F}_x + \vec{F}_y + \vec{F}_z$ that can be represented as

$$\begin{pmatrix} F_x \\ F_y \\ F_z \end{pmatrix} = \begin{pmatrix} M_{xx} & M_{xy} & M_{xz} \\ M_{yx} & M_{yy} & M_{yz} \\ M_{zx} & M_{zy} & M_{zz} \end{pmatrix} \begin{pmatrix} A_x \\ A_y \\ A_z \end{pmatrix}. \tag{1}$$

Subsequently, $\vec{F}_x$, $\vec{F}_y$, and $\vec{F}_z$, are the individual services of a composite service; and are represented as

$$\vec{F}_x = M_{xx}\vec{A}_x + M_{xy}\vec{A}_y + M_{xz}\vec{A}_z$$
$$\vec{F}_y = M_{yx}\vec{A}_x + M_{yy}\vec{A}_y + M_{yz}\vec{A}_z \qquad (2)$$
$$\vec{F}_z = M_{zx}\vec{A}_x + M_{zy}\vec{A}_y + M_{zz}\vec{A}_z .$$

If $\vec{A}$ denotes the syntactic elements of a composite service, then M denotes the semantics of the composite service.

1) Static Composite Services

If we define the processes and the services that realize a composite service during design time, then the resultant composite service is a static composite service. An example is composite services development using BPEL and Web services [9], in which BPEL and the Web services are specified during design time.

2) Dynamic Composite Services

A composite service formation is dynamic if it meets either of the following conditions:

- The processes of a composite service are defined during run time.
- The services that realize a composite service are discovered, selected, and composed during run time.

An example is discussed in [64], in which the processes are generated using Artificial Intelligence (AI) planning technology.

B. Composite Services Development Life Cycle

Fig. 5 provides a schematic representation of the steps involved in composite services development. The continuous lines refer to the forward processes while the dotted lines refer to the feedback process. Every composite services development starts with identifying that a "service need" exists. A service need refers to something that the user requires. The user can be an individual or an enterprise. Table 1 outlines the steps in composite service development.

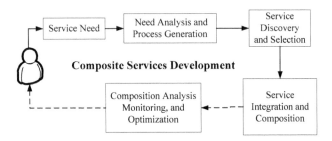

Fig. 5. Composite services development.

TABLE 1
COMPOSITE SERVICES DEVELOPMENT

Steps	Description	Discussion
Need Analysis and Workflow Generation	The first step in composite services development is to analyze and decompose the need into a set of processes.	An approach for specifying processes is using process languages such as business process modeling notation (BPMN), windows workflow foundation (WWF), or business process execution language (BPEL). Formal approaches such as Petri nets, pi-Calculus, and the task system model are also used.
Service Discovery and Selection	The second step is to discover and select the Web services necessary to realize the composite services workflow.	The discovery process can use registries such as the Universal Description, Discovery and Integration (UDDI) [65]. The UDDI specifications define a way to publish and discover information about Web services [66]. The services can be third-party services or the enterprise can develop its own services. Various factors influence the selection of services, such as the quality of the service, security and trustworthiness of the service, the enterprise rules, and business policies [3], [40], [41], [67].
Service Integration and Composition	The next step is to integrate and compose the selected services to form the composite services.	Depending on the process languages used, different composition engines can be used. For instance, BPEL engines can be used if the process is specified in BPEL.
Composition Analysis, Monitoring, and Optimization	Composition analysis and monitoring is required to verify and validate the formation of composite services.	Most composition engines have some approaches for verifying and validating the services. Formal techniques using Petri nets and Pi-Calculus are used for verification and validation (V&V).

C. Composite Services Architecture

In this section, an overview of the different composite services architecture is provided.

1) Service-Oriented Architecture and Web Services

Newcomer and Lomow describe SOA as a style of design that guides all aspects of creating and using business services throughout the development life cycle [41]. The SOA lifecycle runs from the conception of the business service to its retirement. SOA enables enterprises to compose loosely coupled applications from individual services. Web services can be defined as a collection of functions that are packaged as a single entity and published to the network for use by other programs [68]. Web services are building blocks for creating open distributed systems and allow companies and individuals to make their digital assets available worldwide quickly and efficiently. Fig. 6 describes a basic SOA [69]. It consists of three entities: service providers, service brokers, and service requesters. Service providers create services and publish them on service registries of service brokers for service requesters to discover and use. The service registries contain lists of service providers and services. The service providers establish and maintain the service registries.

The following composition example illustrates the roles of service requester, service provider, and a service broker. An airline company (company A) develops a service that lists the schedules of its flights to different destination. It publishes its services on the travel broker's registry (company B). A third company (company C), a travel agent, looks up different airlines in company B's registry, finds the service of company A, and requests its flight schedule for a particular destination and date. In this scenario, company A is the service provider, company B is the service broker, and company C is the service requester. While SOA applications can be composed using non-standardized approaches, Web services provide different standards that allow enterprises to easily adopt SOA [70], [71]. Web services can be said to enable the SOA model of loosely coupled network components in which processes can be composed to form composite services [11], [72]-[76].

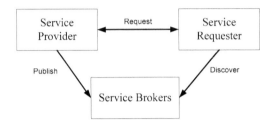

Fig. 6. Basic service-oriented architecture.

Fig. 7 depicts the Web services stack adopted from [72]. The primary layers depicted in the central box of Fig. 7 are for messaging, description, and processes [77]. The other layers depicted at the side boxes of Fig. 7 are for management and security. SOAP is used for exchanging messages between services [72]. It provides a standard, extensible, and composable framework for packaging and exchanging XML messages. Web services description language (WSDL) is used for describing Web services [72]. Service aggregation, orchestration, or choreography is the use of Web services coordinated to achieve a composite goal [78].

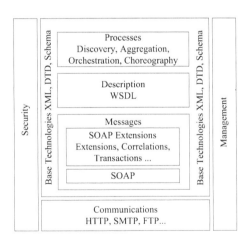

Fig. 7. Web services stack.

2) Semantic Web and Semantic Web Services

The World Wide Web Consortium (W3C) describes the semantic Web as a Web of data [66]. The semantic Web architecture is part of the W3C's effort to improve the current Web with a standardized framework that allows the sharing of data and the reuse of processes across enterprise applications [38], [76], [79]-[81]. The semantic Web is as an extension of the current Web in which we give information well-defined meaning, better enabling computers and people to work in cooperation [82]. The semantic Web is not an application but an infrastructure on which many different applications can be developed [83]. The major efforts in the semantic Web research are the development of new content markup languages, such as the ontology inference layer (OIL), Defense Advanced Research Projects Agency (DARPA) agent markup language (DAML), DAML+OIL, and DAML Services (DAML-S) [74]. These efforts have led to the W3C specifications of the Web ontology language (OWL) and the revised resource description format (RDF) format [74], [84]. OWL facilitates greater machine interpretability of Web content than that supported by extensible mark-up language (XML), RDF, and RDF Schema (RDF-S) by providing additional vocabulary along with a formal semantics. OWL has three increasingly expressive sublanguages: OWL-Lite, OWL-DL, and OWL-Full. The National Science Foundation European Union strategic workshop identifies the four directions of semantic Web development [83]:

- Identification and localization is focused on developing approaches for identifying resources, comparing or equating different identifiers, and localizing Web resources for easier access. This task involves the research of ontology and mark-up languages and developing infrastructure.
- Assessing relationships and reducing differences among semantic models is focused on developing approaches to handle the heterogeneity of the Web. The approaches include developing layered and modular representation languages and metric support.
- Tolerant and safe reasoning is focused on developing tolerant computing techniques to handle the open character of the Web. Additionally, it involves the development of an infrastructure for safe computing.

16

- Facilitating semantic Web adoption is focused on developing supporting infrastructure, such as well-defined ontology libraries, text mining techniques, and unobtrusive collaboration support.

The semantic Web services using the ontology of services OWL services (OWL-S), originally referred to as DAML-S, is an effort to improve the current Web Services model with richer semantic descriptions. OWL-S is developed as part of the DARPA Agent Markup Language (DAML) initiative. OWL-S is an OWL-based Web service ontology that supplies Web service providers with a core set of markup language, constructs for describing the properties and capabilities of their Web services in unambiguous, computer-interpretable form [84]. Fig. 8 depicts the semantic Web services stack.

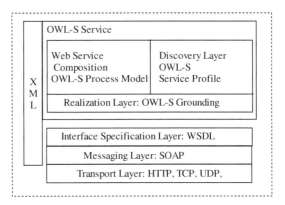

Fig. 8. Semantic Web services stack.

Alesso and Smith list the fundamental automatic OWL-S tasks or the life cycle of the semantic Web services [40]:

- The automatic Web service discovery task involves the automatic discovery of Web services.
- The automatic Web service invocation task involves the automatic execution of an identified Web service by a computer program or agent.

17

- The automatic Web service composition and interoperation task involves the automatic selection, composition, and interoperation of Web services.
- The automatic Web service execution monitoring task involves the monitoring of the execution of individual and composite services.

Fig. 8 depicts the semantic Web stack, showing that it extends on top of the WSDL layer. The extended part of the semantic Web services stack has three layers providing the following knowledge about services [40]:

- The service profile provides a way to describe the services offered by the providers, and the services needed by the requesters.
- The service model describes how a service works.
- The service grounding describes how to access a service.

3) OSGI

The OSGI alliance describes the OSGI service platform as a dynamic Java module [66]. The focus of the OSGI alliance is on standardizing the integration aspects of software to enable efficient and reliable reuse of existing components. OSGI uses composition technology to enable networks to change their composition dynamically. The popular Java-Spring architecture uses OSGI platform as a dynamic service model [66].

The OSGI platform refers to applications as bundles. The platform consists of a number of layers [66]:

- L0: Execution environment layer describes the Java environment specification, such as configuration and profiles.
- L1: Modules layer defines the class loading policies of the environment. The modules layer extends Java with enhanced modularization.
- L2: Life cycle management layer adds bundles that can be dynamically installed, started, stopped, updated, and uninstalled.
- L3: Service registry layer adds a service registry providing a cooperation model that considers the dynamics for bundles.

The OSGI white paper outlines that OSGI platform and provides the following functions [66]:

- The specifications function provides a packaging format to incorporate the bundle in different environments.
- The install a bundle function provides a mechanism to install a bundle in different environments. This function also includes a mechanism to prepare the environment to execute the bundle.
- The start/stop a bundle function provides a mechanism to start or stop a bundle. Starting a bundle makes certain resources available. Stopping a bundle cleans up the resources.
- The update a bundle function provides a mechanism to update a bundle. This function includes mechanisms to stop existing bundles, clean up their resources, unload and replace the unloaded code with new code, and restart the bundle.
- The uninstall a bundle function provides a mechanism to uninstall a bundle, including removing the code and resources from an environment.

4) Other Architectures

Table 2 lists the other composite services architectures.

D. Developing Composite Services with Service-Oriented Architecture and Web Services

In this section, the steps to develop composite services with SOA and Web services are described.

1) Types of Web Services

Web services are divided into two types based on the approach that the services use to transfer messages as representational state transfer (REST) and SOAP-based Web services [94]-[96].

TABLE 2
OTHER COMPOSITE SERVICES ARCHITECTURES

Architecture name	Description
eFlow	The eFlow is a composite services architecture that supports the specification, enactment, and management of composite services [85]. The composite services are modeled as processes, and enacted by a process engine. eFlow also supports a number of features, such as transactions, security, and monitoring.
SHOP2	Simple hierarchical ordered planner (SHOP)2 is a derivative of the SHOP project [86]. SHOP2 is a hierarchical task network (HTN) planning system that can be used for composition. HTN planning is an AI planning methodology that creates plans by tasks decomposition. SHOP2 plans tasks in the same order that they will be executed. By planning for tasks in the order that those tasks will be performed, SHOP 2 makes it possible to know the current state at each step in the planning process [84]. The advantage of knowing the state at each step is that it makes it possible for SHOP2's precondition-evaluation mechanism to incorporate significant reasoning power and the ability to call external programs. This makes SHOP2 ideal as a basis for developing composite services.
UniFrame	The UniFrame project is a composite services framework [87]. The main focus is on the development of the following [88]: A meta-model for components and associated hierarchical set-up for indicating contracts and constraints of the component, an automatic generation of glue and wrappers, based on designer's specifications to achieve interoperability, a formal mechanism for precisely describing the meta-model, and the formalization of the notion of the service quality of each component and an ensemble of components.
Taverna	Taverna is a tool for the composition and enactment of processes [89]-[91]. Taverna is predominantly used in the life sciences domain. Taverna provides a model-driven architecture (MDA) approach for designing processes. The processes are written in a language called simple conceptual unified flow language (Scufl)
ICARIS	Infrastructure for composability at runtime of Internet services (ICARIS) supports three composition techniques [92], [93]: It allows the creation of a service interface, allows stand-alone composite services to be created, and facilities the creation of stand-alone composite services with a single body of code.

REST Web services are based on the REST architecture [95]. They usually use the traditional Web approach of HTTP GET or POST to transfer messages. HTTP GET and POST are used to promote stateless interactions [97].

SOAP-based Web services are of two styles. Document-style is one style of SOAP-based Web services [94]. Another style is the remote procedure call (RPC). The document-style is distinguished by exchanging XML documents as data between Web services. In RPC, the Web service describes the interface in the format of a method signature and takes input and output in a programming-language-specific data type.

The difference between document-style and RPC Web services is provided in the WSDL specification [98]. The WSDL specification currently describes three binding extensions: HTTP GET/POST, Multipurpose Internet Mail Extensions (MIME), and SOAP version 1.1. The SOAP extension allows the style of the SOAP message to be declared as either document or RPC. If the style attribute is declared in the "soap:binding" element, then that style attribute becomes the default for all "soap:operation" elements that do not explicitly declare a style attribute. If the style attribute is not declared in the "soap:binding" element, then the default style is document. The following is an explicit declaration of document-style:

```
<soap:binding style="document" transport="uri">
```

"uri" stands for uniform resource identifier, which is defined as short strings that identify resources in the Web, such as documents, images, downloadable files, and services [99]. Regardless of the declaration within the "soap:binding" element, the "soap:operation" element can override the declaration for each operation:

```
<soap:operation soapAction="uri" style="document">
```

In a SOAP message for which document-style is declared, the message is placed directly into the body portion of the SOAP envelope, as is or encoded. If the style is declared as RPC, the message is enclosed within a wrapper element, with the name of the element taken from the operation name attribute and the namespace taken from the operation namespace attribute.

2) A Contract-First Approach of Developing Web Services

The contract-first approach of developing Web services is also referred to as WSDL-first Web services [42]. Developing Web services with a contract-first approach involves a set of steps. First, the developer analyzes the data sources and designs their data models. Characteristically, this is the most difficult part of developing a Web service. There could be multiple reasons for this, such as a lack of documentation of the data sources as in a legacy situation, or complexity of the data, as in the case of biological sources. After designing the data models, the developer describes the data models, usually using XML schemas. Second, the developer designs the messages that the Web service will exchange. These messages incorporate the XML schemas that were developed in the previous step. Next, the developer designs the operations of the Web service. A WSDL contract is then defined incorporating the operations of the Web services. The operations of the Web service can be of some message pattern, such as request-response or one-way operation. Graphical editors help in the creation of the WSDL. After the WSDL contract of the Web service is developed, the developer can develop the programming stubs, such as Java or C# stubs, from the WSDL interface. These stubs will serve as a guide to the implementation of the Web service.

The contract-first approach is significantly different from the program-first approach. The program-first approach advocates the generation of the WSDL from the programming interfaces. Although the contract-first approach requires greater work upfront and a considerable knowledge of XML to generate the WSDL, its benefits outweigh its disadvantages [98], [100]. Three significant advantages are pointed in [42]: First, the ability of the contract-first approach to support the proper use of XML schemas implies that composite services developers have an improved chance of handling complex data sources effectively. Second, a contract-first approach allows for a more efficient implementation of asynchronous processing than RPC-style approaches as it supports the use of BPEL WSDL extensions. Asynchronous processing is an important requirement of the process-oriented composite services. Otherwise, the processes will need to wait for services that require a long time to complete their tasks. Hence, a contract-first approach can help in improving the reliability, scalability, and performance of the composite service [98]. Third, the approach provides a basis for increasing the granularity of services by enabling services to process documents that naturally fit with the nature of businesses.

22

3) Orchestration or Choreography of Web Services

The differences between the orchestration and choreography models of the composition of Web services is described in [9]. In the orchestration model, a central process controls and coordinates the execution of Web services involved in the operations. The participating Web services are not aware that they are involved in a composite Web service. On the other hand, in choreography, the composition is not dependant on a central operation. Each of the participating Web services is aware of their participation in a composite service. Therefore, the participating Web services know when and how to perform an operation. The orchestration model is currently preferred over the choreography model of composition of Web services because of the following reasons:

- The centralized approach makes the orchestration model easier to manage than the distributed approach of choreography model.
- The orchestration model makes it easier to provide alternative scenarios when faults occur [9].

An excellent comparison of the feature sets of the orchestration and choreography languages is provided in [11]. A popular orchestration language is BPEL. BPEL or BPEL for Web services (BPEL4WS) combine the standards of Web service composition, including Web services flow language (WSFL) and Microsoft's XLANG [89]. BPEL uses a specific set of generic operations to orchestrate Web services. For example, these operations can be of the types invoke, reply, receive, wait, throw (error handling), terminate, and empty (empty operations) [9].

BPEL concepts include both the notion of abstract processes and executable processes. Abstract processes define the business protocol roles while the executable process defines the interaction protocols. Interaction protocols refer to the logic and state of the process that determine the nature and sequence of the interactions conducted by each business partner. BPEL refers to each of the interaction services as partners, which can be both a consumer of service as well as a provider. Partner links define the relationships between partners by defining the message and port types of the interactions. Message types contain both application data and primitive data.

Primitives in BPEL can be activities of the types invoke (invoke a service), reply, receive, wait (wait for some operation to finish), throw (error handling), terminate, and empty (empty operations). Structured activity provides the workflow for the primitives. Structured activity are of the types flow (for parallel execution), sequence (for serial execution), switch (for branching activity), while (for looping activity), and pick (for operations such as timer).

A composition example using BPEL is provided. The example used is a composite service that sends the current weather report to the user via email. The two services needed to realize the composite service are a weather service and an email service. Fig. 9 depicts the workflow of the example composite service. The workflow shows that the user triggers the composition process by providing some input to the system. The orchestration process invokes the weather service and forwards the input of the user. Next, the orchestration process forwards the output of the weather service to the email service. The email service then sends email to the user with the weather report, completing the work of the composite service. For simplicity, assume the user is authenticated, and therefore, the system is aware of the email of the user. The Web service provided by the National Digital Forecast Database is used as the weather service. The email web service was developed using Microsoft .Net. Appendix A provides the implemented BPEL process. I implemented and deployed the example using the Oracle BPEL Process Manager [66].

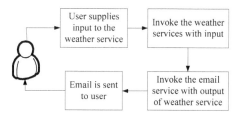

Fig. 9. Workflow for the weather email service example.

E. Introduction to Agents

Agents can provide the processing capability to services in the service-agent model. The processing nature depends on the input, knowledge, and the learning capability of the agents. In certain conditions, agents can simulate human behavior [101]-[103]. In this section, a brief overview of agents is provided.

24

The concept of the software agent evolved from multi-agent systems that form one of three broad areas of distributed artificial intelligence along with distributed problem solving and parallel artificial intelligence [104]. Agents are described in several ways. For instance, [105] describes an agent as a system that, when given a goal, could carry out the details of the appropriate computer operations and could ask and receive advice, offered in human terms, when it was stuck. Nwana describes an agent as a software or hardware that is capable of acting exactingly in order to accomplish tasks on behalf of the user [104]. Agents are described in [106], [107] as follows: To explain the mind, we have to show how minds are built from the mindless stuff, form parts that are much smaller and simpler than anything we would consider smart. However, what could those small particles be - the "agents" that compose our minds?

Agent's research is classified in several ways. In [104], agent research is classified in two strands: The first strand, from 1977 to the present, concentrates on macro issues, such as interaction, communication, and coordination between agents and the second strand, from 1990 to the present, concentrates on a much broader range of agents [104]. Table 3 shows Nwana's classification of agents research [104].

TABLE 3
NWANA'S CLASSIFICATION OF AGENTS RESEARCH

Research Strand	Issues Addressed	Approaches Discussed in Papers
Strand 1	Macro issues such as interaction, communication, and coordination between agents	Bond and Gasser [104] Gasser and Huhns [108] Chaib-draa et al. [109] Gasser et al. [110]
	Theories, architectures, and languages	Wooldridge and Jennings [111] Wooldrige et al. [112]
Strand 2	Diversification in the types of agents being investigated	Nwana [82]

Agents research can also be classified based on two different approaches, agent as an ascription and agent as a description [113]. In the agent as an ascription approach, agents are classified based on the notion of agenthood as an ascription made by some person. In the agent as a description approach, agents are classified based on a description of the attributes that software agents possess. Table 4 shows Bradshaw's classification of agents by Bradshaw [113].

TABLE 4
BRADSHAW'S CLASSIFICATION OF AGENTS RESEARCH

Agent Research Classification	Approaches Discussed in References
Agent as an Ascription	Dennet [114] Franklin and Graesser [115] Kaehler and Patterson [116] Negroponte [117] Singh [118]
Agent as a Description	Shoham [119] Etzioni and Weld [120] Gilbert, et al. [121] Petrie [122] Wooldridge and Jennings [111]

Agents research can also be classified in two ways: traditional AI-based agents and autonomous agents [113]. Table 5 shows the difference between traditional AI research and autonomous agent research [113].

Agents can be characterized in several ways. For instance, agents can be characterized by the concepts of situatedness, autonomy, and flexibility [123]. A classification approach is as follows [124]:

- Based on mobility as static or mobile agents.
- Based on the deliberative thinking paradigm as deliberative or reactive.
- Based on the characteristics of autonomy or cooperation as collaborative agents, learning agents, interface agents, or truly smart agents.
- Based on the agent's roles, such as in the World Wide Web (WWW), as information agents.
- As a hybrid agent, combining two or more philosophies in a single agent.

TABLE 5

DIFFERENCES BETWEEN TRADITIONAL AI AND AUTONOMOUS AGENTS

Traditional AI	Autonomous Agents
Traditional AI focuses on systems that demonstrate isolated and advanced competences. Traditional AI provides depth rather that width in their competences.	Autonomous agents have multiple integrated competences, typically lower level competences. Example of lower level competencies is executing a simple software routine by an agent.
Traditional AI focuses on closed systems that have no direct knowledge of the problem domain about which they encode knowledge and solve problems. The agent interaction with the environment is very controlled often through a human approach.	Autonomous agents are open systems situated in the environment.
Most traditional AI systems deal with one problem at a time.	In autonomous agents, the system is self-contained; that is, the agent figures out by itself the next problem or goal to be addressed.
Traditional AI focuses on the question of what knowledge a system has.	Autonomous agents research focus on the behavior of the system when put in an environment.
Traditional AI is not concerned with the development aspect or the question of how the knowledge structures got there in the first place or changed over time.	The emphasis of autonomous agents research is on the adaptation and on developmental approaches.

Note: Adapted from "Modeling Adaptive Autonomous Agents," by P. Maes, 1994, Artificial Life J., vol. 1, no. 1-2, pp. 135-162. Copyright 1994 by P. Maes. Adapted with permission.

According to the above classification, there are seven major agent types of agents [124]:

- Collaborative agents are agents that emphasize autonomy and cooperation (negotiation) with other agents in order to perform tasks. Collaborative agents might learn, but this is not an emphasized character of their operations.

- Interface agents are agents that emphasize autonomy and learning to perform tasks. Interface agents collaborate with the user in the same environment rather than with other agents.

- Mobile agents are agents that roam the wide area networks (WAN) to obtain information on behalf of their owners. Mobile agents emphasize autonomy and cooperation differently than collaborative agents.
- Information/Internet agents are agents that manage, manipulate, or collate information from distributed sources.
- Reactive agents are agents that simply act or react to the environment in which they are placed.
- Hybrid agents are made up of a combination of two or more agents.
- Smart agents are agents that are truly autonomous, can cooperate with other agents, and learn.
- Heterogeneous agent refers to an integrated set-up of at least two or more agents that belong to different agent classes and including one or more hybrid agents.

Agents are classified into four types based on their architecture [125]:
- Simple reflex agents are agents that select actions based on the current percept, ignoring the rest of the percept history.
- Model-based reflex agents are agents that select actions based on knowledge of the real world.
- Goal-based reflex agents are agents that select actions based on knowledge of the real world and knowledge of the goal.
- Utility-based agents are agents that select actions based on knowledge of the real world and knowledge of the goal and a utility function. The utility function maps a state onto a real number that describes the associated degree of happiness.

F. Conclusion

In this chapter, an introduction to composite services was provided. First, a definition of composite services was provided. The life cycle of composite services development was introduced. The different architectures for composite services development were discussed. An example of process-oriented composite services development with service-oriented architecture (SOA) and Web services was also provided. Finally, a brief discussion of the agent concept used in the service-agent model was provided.

III. PROCESS-ENGINEERING-OF-COMPOSITE-SERVICES-BASED
ARCHITECTURE FRAMEWORK DEVELOPMENT

In this chapter, an overview of process engineering of composite services is provided. Since the term "process" has several different connotations, a definition is provided to clarify its use in this book. The process life cycle that forms an integral part of the composite $P^2_{FRAMEWORK}$ is described. The different process technologies and process formalisms that can be used in process engineering are described. The process-oriented composite services development approaches are classified.

An overview of enterprise architecture frameworks is also provided. The six types of semantics identified in the CPP dimensions and their integration are discussed. The task system model that is used to model the processes of the case study examples is discussed. The process mining approach for analysis of composite services is discussed.

A. *Processes*

The terms workflow, workflow process, and process are used interchangeably in literature and with much confusion. In this section, the use of the term process in this book is clarified.

According to [126], a process is a set of partially ordered steps intended to reach a goal. Any component of a process is a process element. A process step is an atomic action of a process that has no externally visible substructure. Determining that a process element is a process step depends, in part, on whether any further decomposition of the element's structure is needed to support the objectives of the process model.

In the context of manufacturing engineering, a process is considered as not only a physical production of the product but also all of the process that includes the design, marketing, maintenance, and strategic plans of the enterprise [51]. Therefore, process life cycle would include the activities involved in a systematic, structured approach to analyze, improve, control, and manage processes with the aim of improving the quality of products and services.

Sommerville describes a process, specifically a software process, as a set of activities that leads to a production of a software product in [127]. He describes some fundamental activities in the software process management life cycle:

- Software specifications that describe the functionality of the software and constraints on its operations.
- Software design and implementation that ensures that the software meets the specifications.
- Software validation that ensures that the software performs what the customer wants.
- Software evaluation that ensures that the software evolves to meet the changing customer needs.

Delcambre and Tanik describe a process as a total set of activities along with the associated services, tools, methods, structure, and people [32]. They describe the process life cycle as composed of four activities [32]:

- Process modeling, in which the system is designed, models are created, and simulations are performed.
- Process execution, in which the process system is installed and used in the real world.
- Process analysis, in which the process performance is measured during execution and upon reassessment of the goals and execution environment for the process system.
- Process improvement, in which the focus is on the improvement of the performance of the process system.

Performance improvement is relative to its short and long-term goals and upon changes in response to changes in goals and in the execution environment for the process system. The process life cycle described in [32] is similar to the process life cycle described in [128], [129], which is also comprised of four stages: process description and definition, process customization and instantiation, process enactment, and process improvement.

Tanik and Chan describe a process system as a composition of resources and stimuli from the environment, and interrelationship among resources [130]. Functional components transform the data. Control components sequences and coordinates the execution of functional components, response to stimuli, and the transfer of resources to the environment or to other tasks. Therefore, a process can be considered as a potential engineering artifact similar to a software system. In this context, [30] provides a comparison between the software process life cycle of Sommerville and the process life cycle described in [32] and [128]. Whereas the focus of software process life cycles is on requirements analysis and specification, the focus of process life cycles is on explicitly representing and describing the workflow process for composition and analyses. Another difference is that the process life cycles emphasize continuous improvement of the process following enactment in the real world, while the software process life cycles do not explicitly consider this. Table 6 compares the software process life cycle with the process life cycle [30].

TABLE 6
COMPARISON BETWEEN SOFTWARE
AND WORKFLOW PROCESS LIFE CYCLE

Software Process Life Cycle	Workflow Process Life Cycle	
Requirements Analysis and Specification	Process	Modeling
Design	Process Modeling	
Implementation	Process	Modeling
V&V	Process Analysis	
Deployment	Process Execution	
Testing and Maintenance	Process Improvement	

Note: From "A Resource-Focused Framework for Process Engineering," by S.F. Mills, 1995, Ph.D. Dissertation, Dept. School of Engineering and Applied Science, Southern Methodist University. Copyright 1997 by S.F. Mills. Reprinted with permission.

The Workflow Management Coalition (WfMC) defines a workflow process as the automation of a business process, in whole or part, during which documents, information, or tasks are passed from one participant to another for action, according to a set of procedural rules [131]. A business process is defined as a set of one or more linked procedures or activities that collectively realize a business objective or policy goal, normally within the context of an organizational structure defining functional roles and relationships. Therefore, a workflow process management system is a system that defines, creates, and manages the execution of workflows with software, running on one or more workflow engines and is able to interpret the process definition, interact with workflow participants, and, where required, invoke the use of IT tools and applications.

Alternatively, the National Science Foundation (NSF) report on "Workflow and Process Automation in Information Systems" describes a workflow process as an automated organizational process implying automation of the coordination, control, and communication of activities [132]. However, the activities themselves can be automated or performed by people. Workflow process management is the automated coordination, control, and communication of work by both people and computers as it is required to complete an automated process.

VanderAalst and VanHee provide another definition of a workflow process [133], [134]. They describe a workflow process as dealing with cases. The case could be any arbitrary object that needs to be processed in some well-defined way, such as an insurance claim or a patient in a hospital. Generally a business process management system includes workflow management [135]. Therefore, a business process management system can be defined as a system that supports business processes using methods, techniques, and software to design, enact, control, and analyze operational processes involving humans, organizations, applications, documents, and other sources of information. The process life cycle of a business process management system is as follows [135]:

- Design phase, in which the processes are designed.
- Configuration phase, in which the designs are implemented by configuring a process-aware information system, such as a workflow management system.
- Enactment phase, in which the operational business processes are executed.

- Diagnosis phase, in which the operational processes are analyzed to identify problems and to find possible improvements. Most workflow systems offer very little support to the diagnosis and design phase of the process life cycle.

In the context of this book, a process is defined to be made up of the semantic and syntactic elements of composite services. The semantic elements capture the essence of an enterprise, such as representing the work to be accomplished, the role of the user or the user that performs the work, and the services required to accomplish the work. The syntactic elements provide the structure for the process. A process can be modeled using tasks and resources [28]. A task constitutes the unit of compositional activity in a process. The task is specified in terms of its external behavior, such as the input it requires, the output it generates, its action or function, and its execution time. Resources are any (not necessarily physical) device, which is used by tasks. For example, the tasks of a process could use multiples resources, such as a service, software, or a human activity. Process-oriented composite services development emphasises the use of processes throughout the development life cycle. The life cycle of process-oriented composite services is as follows:

- Process modeling, in which the focus is on the development of abstract processes followed by deriving the executable processes from the abstract processes.
- Process composition, in which the focus is on the composition of the executable processes to develop composite services.
- Process analysis and optimization, in which the focus is on the analysis and optimization of the performance of the composite service. Process analysis can also be performed in the process modeling stage. In this case, process enactment would be used to simulate the data for analysis [31].

Fig. 10 shows the life cycle of process-oriented composite services. A mapping of the process life cycle to composite services development is depicted in Fig. 11.

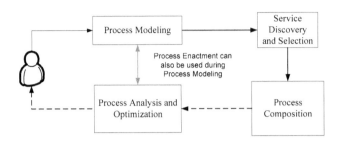

Fig. 10. Composite services development using process-oriented composite services.

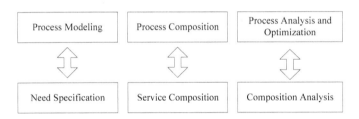

Fig. 11. Mapping of process life cycle to composite services development.

Table 7 compares the life cycle of process-oriented composite services with the Semantic Web service life cycle.

TABLE 7
MAPPING OF COMPOSITE SERVICES LAYER TO
SEMANTIC WEB SERVICE LIFE CYCLE

Semantic Web Service Composition	Process-Oriented Composite Services Life Cycle
Service need analysis	Process modeling
Business process specification	Process modeling
Automatic Web service discovery	Process modeling
Automatic Web service invocation	Process composition
Automatic Web service composition and inter-operation	Process composition
Automatic Web service execution monitoring	Process analysis and optimization

The following steps describe the engineering activity of developing composite services, based on the enterprise model (EM) [13], [22]. Fig. 12 depicts the EM. The process modeling stage starts with the development of abstract process models. Abstract process models should provide support for the semantic and syntactic aspects of an enterprise [8]. The abstract process models provide a good model for process enactment and analysis. The development of the abstract processes starts with a high-level model with few details. The abstract process development continues until models with sufficient details are reached from which it is possible to generate an executable process. The abstract process models are then transformed to executable process models for composition. The executable processes provide an executable representation of the process for composing services. The executable processes must be directly composable on a composition engine and can represent processes that are either mechanistic, that are devoid of any human interaction, or supporting human activities.

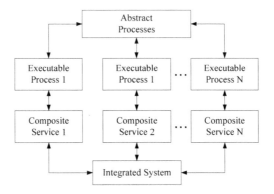

Fig. 12. Enterprise model for composite services development.

The above steps imply that a systematic application of process engineering techniques that can effectively capture the semantic and syntactic aspects of the enterprise becomes an important component of composite services development.

B. Process Modeling of Enterprises

Process modeling is concerned with effectively representing the semantic and the syntactic elements of an enterprise. Semantic aspects capture the intended meaning and behavior of the work, such as when it needs to be completed, who needs to complete it, and what are the necessary resources. Syntactic aspects capture the structure of processes, such as sequential or parallel. Enterprises can be considered as ensembles of the following components [25]: peopleware, composed of people, netware, composed of networks or the communication medium of people, software, which people use for doing their work, and hardware that the software sits on. Accurately capturing the activities of an enterprise requires capturing all of the components of an enterprise. A clearer explanation can be obtained by dissecting the activities entailed in a person doing work in an enterprise; for example, data analysis. The person downloads data from a data source, uses some tools to process and analyze the data, and communicates with his group about the results. A closer examination reveals that the person uses one software to download the data, another software to process and analyze the data, and email or another network program to communicate with his group members. Finally, the software used by the person resides on some hardware, either on his desktop or on the server. Therefore, a process modeling approach can be effective only if it can effectively model the different ensembles of the enterprises and their interactions. Fig. 13 shows that a connection needs to be established between the top (enterprise needs) and the bottom (solutions), in other words, the problem and solution domains.

Tanik and Chan describe an enterprise in terms of data, function, and control [30], [130]. Data and data structures are the objects acted upon or transformed because of the operation of the software system. The operations on the data that describe ways of manipulating data are also a component of data and data structures. Functions provide the capability to transform data; that is, functions perform the real work of the system. Control determines the circumstances under which each function is invoked; therefore, control provides a view into which and when data are processed, and the order of processing.

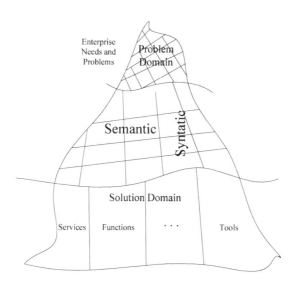

Fig. 13. Process modeling for enterprise development.

Ramamoorthy suggests that an enterprise can be modeled in terms of its functions, features, and control [2]. Functions refer to the activities that are involved in delivering or discharging a user need. Functions can be of four types: manual, manual plus mental, interactive plus mental, and mental activities. Depending upon the type of activity, these require interactions between users, users and services, and between services.

Control can be of four types (Fig. 14) that are as follows:

- Fully isolated, in which each node of a model can represent a team and is independent or isolated.
- Hierarchical, in which the root node of a tree-like structure acts as a controller or director receiving information from its subordinates and sending information to its subordinates.
- Partially decentralized, in which the sub-trees are organized like the hierarchical model. The root nodes of the sub-trees report to the centralized node.
- Fully decentralized, in which there is full connectivity between the teams or nodes. Inter-nodal communication is assumed to be instantaneous.

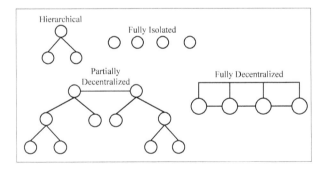

Fig. 14. Control structure of enterprises.

Process modeling approaches should support five activities [136]:

- Facilitate human understanding and communication.
- Support process improvement.

- Support process management.
- Automate process guidance.
- Automate execution support.

Based on the five categories, four perspectives of process modeling are proposed [136]:

- A functional perspective that represents the process elements that are being performed and the flows of information entities, such as data, artifacts, and products that are relevant to these process elements.
- A behavioral perspective that represents when process elements such as sequencing are performed and aspects of how they are performed, such as through feedback loops, iteration, complex decision-making conditions, and entry and exit criteria.
- An organizational perspective that represents where and by whom process elements are performed, the physical communication mechanisms used for the transfer of entities, and the physical media and locations used for storing entities.
- An informational perspective that represents the informational entities produced or manipulated by a process, such as data, artifacts, products, and objects, including both the structure of and the relationship among informational entities.

A process modeling approach must incorporate five sub-models [137]. These sub-models are as follows:

- Activity model that expresses both simple and aggregate activities.
- Product model that expresses data being manipulated by activities.
- Tool model that describes tools and their architecture.
- Organization model that expresses structure and control activities, as well as their behavioral perspective.
- User model that shows how various process actors benefit from the assistance of process technology.

A process modeling approach must incorporate three dimensions [138]:

- An activities dimension concerned with the tasks performed by the human participants of the enterprise.

- A communications dimension concerned with the information exchange between the human participants.

- An infrastructure dimension concerned with the supporting resources required for the execution of processes, as well as the long-term goals of the enterprise.

Table 8 extends Mills [30] comparison of the models described in [130], [136]-[138] with Ramamoorthy's feature, function, and control model [2].

The computer integrated manufacturing open system architecture (CIMOSA) considers process modeling as an activity that involves modeling in three dimensions [139]:

- The derivation dimension is layered into three modeling layers as follows: The requirements definition model captures the business requirements of an enterprise and forms a basis for the derivation of the second and the third layers, the design specification model and implementation description model. The implementation description model includes a computer-executable description of the enterprise operation.

- The generation dimension model deals with the complexity of the problem and the objects to be described. CIMOSA provides four views in the generation dimension: functional, informational, resource, and organizational.

- The instantiation dimension is concerned with the degree of particularization and is divided into three levels: generic, partial, and particular.

Agents are used in process modeling because they can provide a dynamic view of the enterprise characterized by such qualities as situation awareness, mobility, intelligent behavior, and a high degree of distribution [77], [124], [140]. Agent-based modeling approaches can be classified in terms of their methodologies [77]. There are knowledge-based approaches in which agent behavior is dependant on existing knowledge stored in knowledge bases. Examples of knowledge-based approaches are CommonKADS, CoMoMAS, and MAS-CommonKads [141]-[143]. Agent-oriented approaches focuses on the properties of agent systems and try to define a methodology to cope with all aspects of the agent. Example of agent-oriented approaches are gaia and societies in open and distributed agent spaces (SODA) [144]. Recently, a few agent-oriented approaches have focused on extending object-oriented modeling languages such as the unified modeling language (UML) with agent-specific extensions

41

to support the modeling of enterprises. Examples are the model described in [145] and MESSAGE/UML [146].

TABLE 8
COMPARISON OF DIFFERENT ENTERPRISE MODELING VIEWPOINTS

Tanik and Chan [130]	Ramamoorthy [2]	Curtis et al. [136]	Yeh et al. [138]	Conradi et al. [137]
Data and data structures	Features	Informational perspective	Infrastructure dimension	Product model
Function	Function	Functional, organizational perspective	Activities dimension	Activity, tools user model
Control	Control	Behavioral perspective	Communication dimension	Organizational model

Note: Adapted from "A Resource-Focused Framework for Process Engineering," by S.F. Mills, 1995, Ph.D. Dissertation, Dept. School of Engineering and Applied Science, Southern Methodist University. Copyright 1997 by S.F. Mills. Adapted with permission.

C. Process Technology

Process technologies help in process modeling and analysis. In practice, both informal and formal approaches to process modeling and analysis exist. A few of the approaches are reviewed in this section.

1) Model-Driven Architecture:

The Object Management Group (OMG) is the foundation behind MDA. It defines MDA as an approach to system specification that separates the specification of system functionality from the specification of the implementation of that functionality on a specific technology platform [147]. To this end, the MDA defines an architecture for models that provides a set of guidelines for structuring specifications expressed as models. The MDA is as an abstraction model in which the model provides the abstraction. There are four principles that underlie OMG's MDA approach [148], [149]:

42

- Models expressed in a well-defined notation are a cornerstone to system understanding for enterprise-wide solutions.
- Building systems can be organized around a set of models by imposing a series of transformations between models, organized into an architectural framework of layers and transformations.
- A formal underpinning for describing models in a set of metamodels facilitates a meaningful integration and transformation among models, and is the basis for automation through tools.
- Acceptance and broad adoption of MDA requires industry standards to provide openness to consumers and foster competition among vendors.

The OMG divides the development of a system into three model levels and a series of transformation steps to transform the models from one level to another. Fig. 15 depicts the MDA layers. The platform as described in the MDA refers to technological and engineering details. Examples of the platform are CORBA, J2EE, and dotNet. The three models of MDA are as follows [148]-[150]:

- Computation-Independent Model (CIM): OMG describes the CIM as a model that provides a view of the requirements and the environment of the system. The CIM does not provide details of the structure of the system. The CIM role is to bridge the requirements gap between domain experts and software experts.
- Platform-Independent Model (PIM): The PIM captures a view of the system from a platform-independent perspective. The goal of the PIM is to be useful as a model for different platforms of similar types.
- Platform-Specific Model (PSM): The PSM is a view of the system from a platform-specific perspective. A PSM combines the specifications described in PIM with particular details about how it can be used on a particular type of platform.

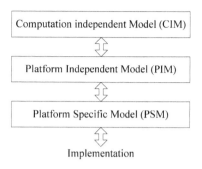

Fig. 15. Process modeling using model-driven architecture.

2) Model-Driven Approach Using Unified Modeling Language

OMG advocates the usage of UML as a modeling language for MDA. It is observed that while UML provides a good language for capturing the structure of a process, it does not have sufficient semantics to capture the behavior and dynamic nature of processes [151]. The use of executable UML (xUML) to overcome this limitation is proposed in [152] and [153]. The xUML model is described as UML without its semantically weak elements, but with the addition of precisely defined action semantics [153]. Therefore, the xUML model is an executable model of UML with the following properties:

- A clearly defined, simple model structure.
- A precise semantics for actions that has been incorporated into the UML standard.
- A compliant action specification language (ACL).
- An accompanying process that is oriented towards executable modeling, large-scale reuse, and pattern-based design.

3) XML Process Definition Language

XPDL is an XML, graph-based process interchange language [154], [155]. The WfMC defines XPDL as a common interchange standard that enables different products to continue arbitrarily supporting internal representations of process defini-

44

tions with an import/export function to map to/from the standard of product boundary [156]. According to WfMC, a process definition language must support the following purposes: act as a template for the creation and control of instances of that process during process enactment, support simulation and forecasting, act as a basis for monitoring and analyzing enacted process, and serve documentation, visualization, and knowledge management. The usability of XPDL to model the flow of Web services is demonstrated in [157]. A critical evaluation of XPDL based on pattern analysis is offered in [158].

4) Business Process Modeling Notation

OMG defines the charter of BPMN as follows [66]: A standard BPMN will provide businesses with the capability of understanding their internal business procedures in a graphical notation and will give organizations the ability to communicate these procedures in a standard manner. The BPMN graphical notation will facilitate the understanding of the performance collaborations and business transactions between organizations. The graphical notation will also ensure that businesses will understand themselves and participants in their business and will enable organizations to adjust to new internal and B2B business circumstances quickly.

The business process management initiative (BPMI) developed the BPMN to bridge the gap between business process design and process implementation [159]. The business process diagram (BPD) of BPMN is based on a flowcharting technique in which the business process model is a network of graphical objects, which are activities and the flow controls that define their order of performance. The BPMN specifications indicate that the scope of the BPMN is restricted to business processes. BPMN is part of the BPM stack, as depicted in Fig. 16 [66].

The BPM stack leverages Web services standards such as the Choreography Description Language and BPEL. The parts of the stack that are part of the BPM specifications are defined as follows [52], [160]:

- Business process modeling language (BPML) is an XML language that encodes the flow of a business process in a form that can be interpreted by the process engine.

45

- Business process query language (BPQL) is a standardized administration and monitoring query language for business processes. The BPM is intended to support business activity monitoring (BAM).

- Business process semantic model (BPSM) provides a semantic model for business processes. The BPSM is defined using OMG meta-object facility (MOF).

- Business process extension layers provide a standard set of BPEL extensions. The extensions are defined using BPEL's standard extension mechanisms.

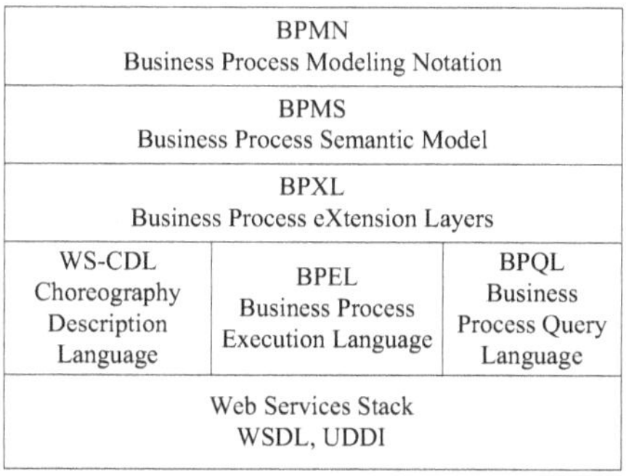

BPMN
Business Process Modeling Notation

BPMS
Business Process Semantic Model

BPXL
Business Process eXtension Layers

WS-CDL Choreography Description Language	BPEL Business Process Execution Language	BPQL Business Process Query Language

Web Services Stack WSDL, UDDI

Fig. 16. Business process modeling stack.

5) Choreography Description Language

The Web services choreography specification is aimed at enabling process designers to precisely describe collaborations between any type of party regardless of the supporting platform or programming model used by the implementation of the hosting environment [161]. The W3C working draft describes the WS-CDL as an XML-based language that describes peer-to-peer collaborations of parties by defining, from a global viewpoint, their common and complementary observable behavior where ordered message exchanges result in accomplishing a common business goal [161]. The goal of the W3C is that WS-CDL must be used in conjunction with BPEL to develop composite Web services [162].

46

6) Business Process Execution Language

BPEL is considered as the de facto standard for orchestrating Web Services [9], [42]. An example of composite services development using BPEL is provided in Chapter II. Techniques to convert UML, BPMN, and WS-CDL to BPEL for composition are provided in [162]-[164].

7) Windows Workflow Foundation

The WWF is a general purpose programming framework for creating reactive programs that act in response to stimuli from external entities [165]. It is a combination of a programming model, engine, and tools for developing workflow-enabled applications on Windows [166]. The WWF consists of a .NET Framework version 3.0 (formerly WinFX) namespace, an in-process workflow engine.

WWF can be seen as a composite service of activities and supports two types of modeling types: sequential workflow and state machine workflow [167]. Sequential workflow is composed of activities in a sequential flow. State machine workflow is composed of activities initiated by state transitions. Harvey provides a list of the different business process modeling standards and their intended purpose [10], [52]. A feature comparison of the various process-modeling technologies is provided in [11].

D. Process Formalisms

Traditionally, formal approaches such as Petri nets and pi-calculus are used in process modeling and analysis. A brief introduction to some of the approaches is provided in this section.

1) Petri Nets

One of the more popular approaches of process modeling and analysis is Petri nets. Carl Adam Petri first developed Petri nets in his dissertation [168]. Petri nets is a graphical and mathematical modeling tool that can be used to describe processes that are concurrent, distributed, parallel, nondeterministic, and/or stochastic [169]. Peterson describes Petri nets as capable of describing both the static and dynamic properties of a system [170].

In Petri nets, graph-based models are used to describe the static properties of a system. The graph contains two types of nodes. Circles depict places and bars depict transitions. Directed arcs from places to transitions and transitions to places connect the places and bars. For instance, if an arc is directed from node i to node j, then i is an input to j, and j is an output of i.

The dynamic properties of a system can be expressed by executing the Petri nets. The position and movement of markers control the execution of Petri nets. The markers are called tokens and are indicated by black dots. Tokens reside in the circles representing the places of the net. Petri nets with tokens are called marked Petri nets.

In literature, several variations of Petri nets are used in modeling and analyzing processes. An example is the FUNSOFT nets [30], [171]. Petri nets also form a basis for informal process modeling technologies. An example is the Yet Another Workflow Language (YAWL), a Petri net-based approach designed to model processes [43]. Other examples of process modeling technologies with a Petri nets basis are BPMN and WSFL [52].

2) Pi-Calculus

A discussion of the pi-calculus approach was first presented in [172]. Later, Miller also presented a tutorial in which he describes the different aspects of pi-calculus [173]. Pi-calculus is an algebra-based formal language for describing concurrent processes including but not restricted to business processes [52].

Parrow describes pi-calculus as a mathematical model of processes whose interconnections change as they interact [174], [175]. The basic computation step is the transfer of a communication link between the two processes. The recipient can thus use the link for further interaction between the two parties. This makes pi-calculus a suitable approach for modeling systems where the accessible resources vary over time.

Harvey observes that pi-calculus, despite its academic background, has become a basis for several process-modeling technologies, such as BPEL, BPML, and WS-CDL [52].

3) Control and Cubic Flow Graphs

Flow graphs provide a graph-based formal technique for process modeling and analysis. Flow graphs provide an underpinning of graphical theoretical notion to process analysis [176]-[178].

Allen defines a control flow graph as a directed graph in which the nodes represent basic blocks and the edges represent control flow paths [179]. A control flow has two distinguished nodes, the entry node and the exit node [176]. An imaginary node originates from the exit node and terminates at the entry node to make the graph strongly connected. As the control flowgraphs are strongly connected, any two nodes are mutually reachable. There are three classes of nodes in the control flow graphs: decision-making nodes, junction nodes, and process nodes.

Cubic flowgraphs are a specific class of control flowgraphs [176]. Cubic flowgraphs can be used effectively to analyze complex processes because of the following reasons [176]: First, the structure of processes can be modeled using cubic flowgraphs. Second, the cubic flowgraphs can be decomposed into its sub-units. Then, each sub-unit can be analyzed independently to understand the working of the process. The use of control and cubic flow graphs for modeling and analyzing process systems is described in [180] and [181].

4) Task System Model

Coffman and Dening originally developed the task system model for representing and analyzing the interaction of concurrent software processes within the context of computer operating systems [28]. The task system model is adopted as a process modeling and analysis framework in [29]-[32], [182]. The task system model is designed to represent the following important characteristics of an operating system:

- Concurrency, which is the existence or potential existence of simultaneous parallel activities.
- Automatic resource allocation, which provides a mechanism for automatically allocating the resources of a system.
- Sharing, which allows for simultaneous use of resources.
- Remote access, which provides conversation access to system resources.

- Asynchronous operation, which handles the unpredictability in the order of occurrence of events.

Table 9 shows that the modeling needs of an operating system are comparable to the modeling needs of composite services system. The task system model provides support for analyzing and identifying process problems, such as determinacy, deadlocks, mutual exclusion, and synchronization. The task system model is used as an abstract processes modeling approach in this book. A description of the task system model is provided in Section J.

TABLE 9
FEASIBILITY OF USING TASK SYSTEM MODELING
FOR COMPOSITE SERVICES

Operating System Characteristics	Composite Services Characteristics
Concurrency	The process in composite services development establishes the order of composition.
Automatic resource allocation	Once a process is established, the next step is to discover and compose the services. If a service is busy with another request, the composite services system might have to wait until that service is free or use another service that performs the same task.
Sharing	Two or more processes of a composite services system might share the same service for its operation. Therefore, a system should be in place for handling multiple requests to a service. In a system in which the development of a service might not be in the control of the enterprise, the process should be able to handle sharing of resources.
Asynchronous operation	In some cases, when the execution of service is time consuming, asynchronous operation plays an import role in ensuring the efficient operation of a system. Checks must also be in place to handle the potential failure of a service. That is, the process should have mechanisms such as timeout to ensure that the system does not hang waiting for a failed service.

E. Classification Model of Process Formalisms

Process modeling formalisms can be classified as descriptive, network-based, imperative or programmatic, and hybrids [30], [31], [130]. Table 10 summarizes the classification details [30].

TABLE 10
CLASSIFICATION OF PROCESS MODELING FORMALISMS

Classification Type	Description	Examples
Descriptive	Descriptive systems provide a framework for characterizing the key entities and relationships in the process. An advantage of descriptive systems is that they do not require an explicit ordering of tasks. As a result, they allow for inherent parallelism. Descriptive systems also do not require explicitly defining the order sequence [183]. Therefore, the ordering of tasks is determined during run time. This property allows for easily modifying the behavior of the system by changing the input to the system. However, this also causes a disadvantage in that we cannot easily visualize the process model and execution.	Rule-based systems are examples of descriptive systems [125].
Network-based	Network-based formalisms provide a framework for modeling of non-determinism and parallelism typically as a graphical view of the system. The graphical view allows for clearly identifying the sequence and concurrency of the process. Network-based formalisms have certain disadvantages: They can become unwieldy and difficult to modify when modeling large processes. The nature of network-based formalisms can often lead to conceptual problems when modeling complex processes [184]. Disruptions can occur when making changes dynamically to processes [183].	Petri nets and finite state automata are examples of network-based formalisms.
Imperative or programmatic	Imperative or programmatic formalisms are modeling approaches that are similar to programming languages. The advantage of imperative or programmatic approaches is that they can easily be executed on machines. The disadvantage is that these approaches do not inherently support non-determinism.	An example is the Ada Process-Programming Language, which is now an obsolete approach [185]. Another example is the JIL process programming language [186].
Hybrids	Hybrid approaches are combinations of two or more of the above techniques.	An example is the task system model.

Note: Adapted from "A Resource-Focused Framework for Process Engineering," by S.F. Mills, 1995, Ph.D. Dissertation, Dept. School of Engineering and Applied Science, Southern Methodist University. Copyright 1997 by S.F. Mills. Adapted with permission.

F. Types of Process Modeling Approaches

Processes of an enterprise can be modeled using different approaches. Four types of modeling approaches are discussed in this section: activity models, state machine models, goal-oriented models, and role-activity models. An example of the four modeling approaches is provided using the weather email service example introduced in Chapter II.

Activity models are based on decomposing the processes into activities [187]-[190]. UML activity models are a popular technique. Fig. 17 depicts the weather email service example using the UML activity modeling diagrams. Activity models are very popular because of its simplicity. In state machine modeling, the states of the tasks of a process and their transitions are modeled [191], [192]. Transitions in a state machine occur because of some events. Fig. 18 represents a state machine model of the weather example using the UML state chart modeling diagrams. State machine models are used typically when a process has several events occurring in it. Goal-oriented models are used typically to model the goal-oriented behavior of the agents. Goal-oriented modeling is achieved by decomposing the process based on goals. A goal could be defined as a non-operational objective (e.g. users, or external systems), to be achieved by the composite system [193]. Goal-oriented modeling is described in [194], [195]. Fig. 19 depicts the weather example using a goal-oriented modeling approach. Role activity modeling is a modeling technique that organizes process based on roles [196]. Role activity modeling is achieved by decomposing a process based on the roles of the users of the system [34], [197]. Fig. 20 represents the weather example using role activity models.

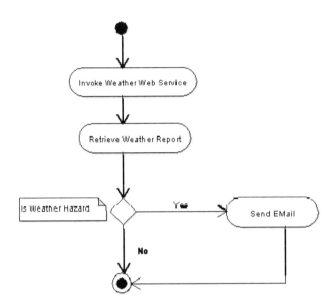

Fig. 17. Activity modeling.

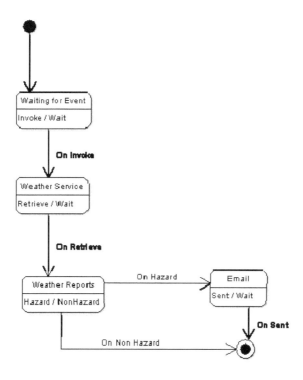

Fig. 18. State chart modeling.

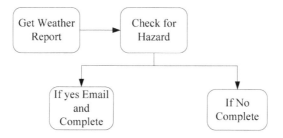

Fig. 19. Goal-oriented modeling.

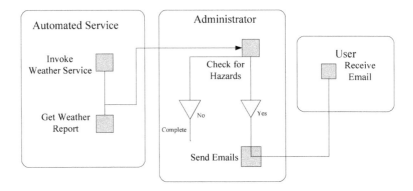

Fig. 20. Role activity modeling.

G. Process-Oriented Composite Services

This section provides a review of composite services approaches. As discussed before in Chapter II, composite services development can be divided into four stages: need analysis and workflow generation, service discovery and selection, service integration and composition, and composition analysis, monitoring, and optimization. Once a need is established, several technical issues should be addressed to ensure the effective development of composite services. In [198], four important requirements of composite services architecture are described:

- Coordination to control the execution of a composite service.
- Monitoring of events that happen during the execution of a composite service.
- Conformance, that is, a requirement to ensure the integrity of the composite service by matching its parameter types with those of its components, impose constraints on the component services, and perform data fusion activities.
- QoS composition that is required to leverage, aggregate, and bundle the component's QoS to derive the composite QoS, including cost, performance security, authentication, privacy, integrity, reliability, scalability, and availability.

A list of issues in composite services development is also presented in [199]. Table 11 compares the issues presented in [199] and [198]. A comprehensive overview of current solutions to the composite service development is presented in [200]. A survey of agent-based approaches for composite services development is presented in [201].

Process-oriented composite services efforts can be classified as non-agent-based approaches and agent-based approaches. Non-agent efforts use a variety of means to describe processes. They can either use process technologies, such as UML or BPEL, or process formalisms, such as Petri nets, or patterns of processes for composition. Hybrid efforts that use a combination also exist. Fig. 21 depicts our classification model. Table 12 provides a classification of the process-oriented research and the associated research efforts.

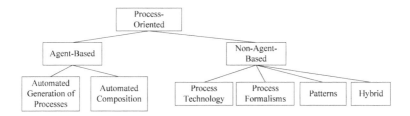

Fig. 21. Classification of process-oriented composite services development.

TABLE 11

ISSUES IN PROCESS-ORIENTED COMPOSITE SERVICES

Dustdar and Schreine [198], [199]			Papazoglou and Geor-gakopoulos [198]
Issue	Description	Efforts	
Coordination	Composite services require coordination and control of sequences.	Ongoing standardization efforts aim to address this problem [72].	Coordination
Transaction	Transaction support to guarantee the interactions of services is necessary during their composition. Short- and long-term transactions should be supported [198].	Example of an ongoing efforts is WS-Transaction [197].	QoS Composition
Context	Context is important because it provides information to adjust execution and output to provide the client with a customized and personalized behavior [202], [203].	Kepler et al. and Alvarez et al. [202], [203].	Conformance in part can be handled by the context of services.
Conversation Modeling	Conversations between services are a useful means of structuring communicative interactions among agents [204].	The conversation of Web services can be modeled using Petri nets to aid their composition [204].	QoS Composition
Execution Monitoring	Dustdar and Schreine discuss two types of execution of composite services: centralized, that is, similar to the client-server, and distributed which expects the participating Web services to share their execution context [199]. The two models are similar to the orchestration and choreography models.	The examples provided by Dustdar and Schreine are eFlow for centralized and SELF-SERV for distributed model [205] .	Monitoring
Infrastructure	Infrastructure provides QoS constraints for discovery of services [27].	Ran et al. model of discovery of services [27].	QoS Composition

TABLE 12

CLASSIFICATION OF PROCESS-ORIENTED COMPOSITE SERVICES
RESEARCH

Classification Type	Description	Approaches Discussed in
Non-Agent-Based	Process description using process technology	Skogan et al. [206], de Knikker et al. [94], Suvee et al. [207], Albert et al. [208], and Ludscher et al. [209]
	Process description using process formalisms	Pankratius and Stucky [210], Thomas et al. [211], and Tan et al. [212]
	Pattern-based process description	Benatallah et al. [213], Everaars et al. [214], Charfi et al [215], and Tut et al. [216]
	Hybrid efforts	Yacoub and Ammar [217], Dong et al. [218]
Agent-Based	Automated generation of processes	Doshi et al. [64], Klusch et al. [219], Qiu et al. [220], and Gekas and Fasli [221]
	Automated composition-focused	Mandell and Mcilraith [4], Sivashanmugam et al. [222], Nie et al. [223], Yiu et al. [224],Tsai et al. [225], Lammermann and Tyugu [226], Thakker et al. [227], Cao [87], and Chen et al. [228]

An example of composite services development using MDA and UML is provided in [206]. They describe a four-step model in which the focus is on the use of the activity diagrams of UML to design the composition. An important step in this model is the transformation of WSDL to UML. Once the UML design is developed, they can be transformed into any executable process. The composition of services using the BPEL executable processes (transformed from UML) and the WorkSco composition engine is demonstrated in [206].

The use of BPEL to compose different bioinformatics services is demonstrated in [94]. The use of BPEL for composition is compared in [94] with a hard-coded application in Java and another workflow technology predominantly used in bioinformatics called Taverna [89], [90].

Aspect-oriented software development (AOSD) is used for composite services development in [207]. AOSD programming technologies (aspect-oriented programming) provide linguistic mechanisms for separate expression of concerns, along with implementation technologies for weaving these separate concerns into working systems [229].

The use of FuseJ, an architectural description language, for unifying aspects and components as an approach for the composition of Web services is described in [207]. In FuseJ, two Web services can be composed in an aspectual manner that is triggered as an advice. The same functionality can also be integrated in a third Web service in a non-aspectual manner.

A Petri nets-based formal algebraic notation is developed to demonstrate the use of Petri nets for composite services development [210]. The notation is used to demonstrate the creation and composition of Web services. The flow of messages and methods in a Web services transactions is modeled using Petri nets in [211]. The Petri-net models were developed as an aid to ensure the absence of deadlock and the correct termination of the transaction. An automated tool to generate the Petri-net model from the Web service description is also developed in [211]. The use of Petri nets as a guide to process-oriented composite services development is demonstrated in [212].

Interaction patterns that depict processes can also be used in composite services development. An interaction pattern can be described as a three-part rule that expresses a relation between a certain context, a problem, and a solution [230]. Tut and Edmond investigate the use of patterns as a composition approach. The authors describe a series of patterns and discuss their use for composition. Another example of pattern use in composite services development is discussed in [213], which presents patterns as a means of realizing bilateral as well as multilateral execution of composite services.

Patterns can be described using process technologies such as UML or process formalisms such as Petri nets, demonstrating the value of hybrid approaches. Yacoub and Ammar provide an example of using UML for pattern-based composition [217]. UML provides an MDA approach of describing patterns at various levels of abstraction. Yacoub and Ammar describe a pattern-oriented analysis and design (POAD) technique. In POAD, the patterns are described at three increasing levels of detail.

The three levels are the pattern level, the pattern interface level, and the pattern detailed level. The three levels are used for analysis of the composition.

Another example of a hybrid model is using process formalisms to analyze process descriptions specified using a process technology. Petri nets is used to test and analyze BPEL-based composite services development in [218]. A translation mechanism to convert BPEL-based definitions to high-level Petri nets is demonstrated. The high-level Petri nets are then verified on existing mature tools and the related researches on high-level Petri nets.

The focus of agent-based efforts is on automation. The agent-based efforts can be sub-classified as approaches that focus on automating the generation of the processes and approaches that focus on automating the service composition after the design of the processes. The processes can be described using one of the non-agent approaches.

The use of Markov chains to automate the generation of the processes is demonstrated in [64]. Using Markov chains allows the resultant processes to incorporate non-deterministic behavior and adapt to the dynamic environment. The resultant processes are generated at an abstract level. A semantic Web service planner called OWL-S-XPlan is described in [219]. OWL-S-XPlan is designed to allow fast, flexible development of composite services in the semantic Web services architecture. From the semantic Web services description and the domain description, OWL-S-XPlan generates the process plan sequence to satisfy a given goal. OWL-S-XPlan extends an action-based FastForward planner with a HTN planning and replanning component.

The use of BPEL combined with the power of automating the composition through OWL-S is described in [4]. In their model [4], the authors describe the executable process using BPEL, and develop a semantic discovery service that will use the BPEL description to automate the discovery and composition of services that are described using OWL-S.

Semantic process templates (SPT) is used to capture the semantic requirements of the process in [222]. SPT acts as configurable modules allowing the maintenance of the semantics of the participating activities, such as control flow, intermediate calculations, and conditional branches, and exposes the semantics in an industry-accepted interface. The templates are instantiated to form executable processes according to the semantics of the activities in the templates.

60

H. Nondeterminism

Nondeterminism is described as follows [30]: when more than one event is possible at some point during an execution of a process, the system must make a choice between the events. Deterministic behavior occurs if the choice between the events is based entirely upon environmental factors. Nondeterministic behavior occurs if the selection among alternatives is made in such a way that the environment does not entirely influence the selection. Nondeterminism is classified as two types: local and global [231]. Local nondeterminism occurs when a process P_i can communicate with any of $P_{i_1} ... P_{i_n}$ and decide on its own which communication to wait, independent of any consultation with other processes [232]. Global nondeterminism is resolved by inspecting the willingness of other processes to communicate. Only mutual willingness to communicate may result in de facto communication.

Nondeterminism is unavoidable in a process with human interaction because of the way people work. People work in the REACT pattern [34]. REACT stands for research, evaluate, analyze, constrain, and task. That is, when people need to do some work, they often research it before taking action. Research can take several forms. The person can access online information, go to a library, or collaborate with others. Once a person gains more information, they evaluate and analyze the information to make a decision. Constrain refers to the process of breaking a job into smaller chunks to simplify it. Task means allocating the smaller chunks to appropriate people. As a result, a person often makes a decision based on new knowledge that affects the nature of the process.

Processes cannot easily account for this nondeterminism [30]. For instance, process escalation is a form of nondeterminism [163] that occurs because we design the process model based on our knowledge of an organization's objectives, infrastructure, context, and constraints. At run time, this idealized view is often broken. In particular, process models generally assume that planned activities happen within a certain period. When we make such assumptions, users must make decisions regarding alternative arrangements to achieve the goal of completing the process within its expected period or to minimize tardiness.

It is not possible to implement systems fully accounting for nondeterminism [232], [233]. It is a myth that technology can supplant the unreliable human [55]. We can only address the problem by implementing a system that can reasonably address nondeterminism [234]. Process models typically account for nondeterminism as an asynchronous task. This implies that the other parts of the process execution can proceed without waiting for the task to complete. The asynchronous task then performs a callback to send its response. Checks and balances are placed to handle nondeterminism. For example, time out is used in case the task is not executed in a reasonable amount of time. A few approaches to handling nondeterminism are using other resources to change the routing of work, changing the work distribution, or changing the requirements with respect to available data.

A CPP framework, such as the composite $P^2_{FRAMEWORK}$ can only enhance the ability of enterprises to handle nondeterministic behavior. For instance, because the CPP semantics are added, it is easier for process analysts to understand and optimize the processes. A description of the CPP semantics is provided in Chapter IV. Processes that are aware of the semantics of user needs might be better equipped to handle nondeterminism because these processes can react to nondeterminism more competently. The composite $P^2_{FRAMEWORK}$ also supports the development of flexible composite services. If a process is continually underperforming, then the process must be identified, studied, and optimized. Finally, the composite $P^2_{FRAMEWORK}$ advocates automation, which can reduce the nondeterministic behavior in enterprises.

I. Characteristics of Human Interactions with Processes

A survey was conducted that shows the importance of CPP for team effectiveness [235]. The study reiterates that understanding processes interactions with humans should be an important consideration of enterprise integration. The link to human interactions is the missing link in process management tools [34]. Both [235] and [34] suggest that without proper support for CPP, the management and analysis of processes are also ineffective because they operate based on an incomplete view of the system's process.

A process-engineering approach should consider the following for supporting human interactions [34]:

- Connection visibility that provides a strong representation of process participants, the roles they play, and the private information resources that belong to each of them,

62

- Structured messaging that provides a process context for the interaction between different users,

- Support for mental work that recognizes the value of information processing done in people's head, and offers a way to manage and recompense mental work like other forms of activity,

- Supportive rather than prescriptive activity management that supports the nature of human work rather than prescribing it, and

- Process change processes that must be able to effect continual change to the process itself.

Four types of distributions of teams in an enterprise is described as follows in [236]:

- Spatial distribution: Teams can be located at geographically distributed locations.

- Temporal distribution: In several cases, teams and individuals interacting with a process can change.

- Technological distribution: Teams use sophisticated tools to interact with processes.

- Social distribution: Teams play different roles with different expertise in a process.

The need for efficient management of the tools and resources of users, and the management of the interaction of the users in a process is also stressed. The importance of tools and resource management is stressed in [237]. The authors observe that the information needed by users to interact with processes is stored in different tools. For example, users manage their information using such tools as email, documents, and browsers. The information should be integrated efficiently to ensure that users can accomplish tasks effectively and on time. A model to ensure that the communications between team members can be tracked effectively is discussed in [238]. The tracking of communications allows the team to make appropriate decisions.

Techniques for efficiently managing the interactions of users and teams in a process is discussed in [239]. The development of a dynamic interaction generation team system for supporting of interactions of humans with machines using automated reasoning is also discussed in [239]. The importance of supporting human interaction anywhere and anytime is discussed in [240]. The authors demonstrate an open framework that enables enterprises to design an interaction system that provides enhanced user support and easy integration of new devices.

The importance of monitoring user interaction with machines to optimize and identify integration needs of the user is discussed in [241]. After studying the features of service-driven enterprises, Ramamoorthy identifies five distinguishing characteristics of service functions as listed in Table 13 [2].

TABLE 13
SERVICE FUNCTION CHARACTERISITICS

Features	Description
Human-needs-driven	This feature incorporates the notion of humanization and individualization. Humanization is the process of supporting the need to make the machines react more like humans so that the operator or user is always comfortable. Individualization is the process of customizing the application to individual needs.
Knowledge-intensive, high mental support	As humans understand their service process better, they demand better, faster, and more intelligent service interactions. Enterprise systems must be able to adapt to these changes
Automation-intensive to reduce manual effort	Automation refers to replacing human interaction with automated system whenever appropriate. As we better understand service interactions, we can optimize them with automation when necessary.
Human-interaction-intensive	Human-interaction-intensive feature incorporates two types of interactions of humans with machines. The first is a simple interaction in which humans are simply involved in the transactions, such as transferring data. They are not involved in any decision making. The second type of interaction is a knowledge-intensive interaction in which humans are involved in the decision-making processes that require mental effort.
Information-technology-intensive	This refers to the type of human interaction in which they are heavily dependent on tools and technology in their work.
Team-based	This refers to the different types of team interactions. Teams could be isolated, hierarchical, or partially or fully decentralized.

These papers indicate that a CPP framework should provide guidance for developing composite services that addresses the following issues:

- Integrated information: Integrated information is needed by users for interacting with processes in an informed manner. This need requires that composite services with CPP must be aware of all of the user's resources and the user's interactions with those resources.

- Different participant roles: Process participants can play different roles in different scenarios. For example, a "faculty" user could play the roles of faculty member or university employee. Therefore, composite services with CPP must incorporate the context of the process and the type of user interacting with the process for that context.

- Flexible and changing processes: As humans gain knowledge about their domain, they continually change their processes to make them more efficient. Composite services with CPP must consider that processes continually change and must be flexible to accommodate the changes.

- Different types of interactions: A human interaction with a process can be a simple interaction or a knowledge-intensive effort. The knowledge-intensive efforts could be a collaborative process involving several people and several information technology tools. Composite services with CPP must be aware of the type of interactions between the user and the process to support the way the user works.

- Different characteristics of participants: People participating in a collaborative process can be from different localities, and hence the efficiencies of their interactions can be different based on their locality. For instance, a team from the United States might be collaborating with a team from India. Composite services with CPP must recognize that the teams can have different characteristics, such as working in different time zones, and that communication between them is usually not synchronous.

- Dynamic nature of human interaction with processes: The nature of user interactions with process can be dynamic. A user, for example, might be on sick leave and unable to play his part in the process on time. The process cannot simply start from scratch. Support for dynamic nature of human interaction is especially important in processes that operate for several days. Composite services with CPP must support a fail-safe mechanism to handle the dynamic nature of human interaction with processes.

- Automation of interactions: Certain human interactions with processes can be automated using technology. Composite services with CPP must be automated whenever possible. Appropriate areas of automation can be identified to reduce user interactions to improve user experiences and the efficiency of the process.

J. Enterprise Architecture Framework Development

In this section, an overview of enterprise architecture frameworks is provided. The six types of semantics identified in the CPP dimensions and their integration are discussed.

1) Enterprise Architecture Framework

The following definitions are adopted in the composite $P^2_{FRAMEWORK}$ development.

- An enterprise architecture frameworks supports and guides organizations during the development of enterprise architecture [242]. The development process can include system planning, design, building, deployment, and maintenance.
- An enterprise architecture is a strategic information asset base that defines the mission, the information necessary to perform the mission, and the transitional processes for implementing new technologies in response to the changing mission needs [243].
- An enterprise is any collection of organizations that has a common set of goals and/or a single bottom line [244]. The term enterprise can be used to denote both an entire enterprise, encompassing all of its information systems, and a specific domain within the enterprise.

An enterprise architecture should provide the following features [243]:

- Alignment ensures the implemented enterprise meets with the management's intent.
- Integration ensures that the business rules are consistent across organization, that data and its use are immutable, interfaces and information flow

66

are standardized, and connectivity and interoperability are managed across the enterprise.

- Change facilitates and manages any aspect of the enterprise.
- Time to market reduces systems development, application generations, modernization time frames, and resource requirements.
- Convergence strives towards a standard information technology portfolio contained in the technical manager model.

There are several enterprise architecture frameworks in the literature. These include the Zachman framework, The Open Group architecture framework (TOGAF), the extended enterprise architecture framework (EEAF), and the federal enterprise architecture framework (FEAF) [245]. The Zachman framework is described as one of the most comprehensive frameworks and, therefore, many of the other frameworks are based on it [242]. Fig. 22 depicts the layers of the Zachman framework [246]. The "x" in Fig. 22 indicates that the column dimension should be addressed.

	Data	Network	People	Time	Motivation
Objectives/Scope	x	x	x	x	x
Business Owner's View	x	x	x	x	x
Architect's View	x	x	x	x	x
Technology Designer's View	x	x	x	x	x
Builder's View	x	x	x	x	x
Functioning System	x	x	x	x	x

Fig. 22. Zachman framework.

John Zachman introduced the Zachman framework in 1987 [247]. The descriptions of the layers are as follows:

- The scope layer provides a ballpark view of the enterprise. It defines the direction and business purpose of the enterprise.
- The business owner's view layer provides an owner's view of the enterprise. This view defines the business terms of the enterprise, such as the nature of the business, including its structure, functions, and organization.
- The architect's view layer provides a model of the information system. This view adds an information perspective to the business owner's view,

such as the information that needs to be collected and maintained. The architect's view begins to describe that information.

- The technology designer's view layer provides a description of the use of technology needed to address the information processing needs identified in the architect's view.
- The builder's view layer provides a detailed description of the program listings, database structures, and networks.
- The functioning system layer is the implemented system.

Each of the layers consists of columns that describe the dimensions of the system's development effort. They are as follows:

- The data column addresses understanding and dealing with an enterprise's data.
- The function column describes the process of translating the mission of the enterprise into successively more detailed definitions of its operations.
- The network column addresses the geographical distribution of the enterprise's activities.
- The people column addresses the people involved in the business.
- The time column describes the effects of time in an enterprise.
- The motivation column describes the translation of business goals and strategies into specific ends and means.

The EEAF is based on Zachman framework [248]. It adds two columns to the Zachman framework, the with/who and with/what columns. The with/who column describes the collaborating entities of the enterprises. The with/what column provides a solution representation of the enterprise.

The Open Group is developing the TOGAF architecture framework to be an industry-standard architecture method that can be used for developing products associated with any recognized enterprise architecture framework [244], [249], [250]. The TOGAF is adopted from the Department of Defense architecture framework. It consists of ten phases in its life cycle: preliminary phase, architecture vision, business architecture, information system architecture, technology architecture, opportunities and solutions, migration planning, implementation governance, architecture change management, and re-cycle back to architecture vision. A detailed description of the TOGAF framework is provided in [251].

The FEAF is developed by the United States Office of Management and Budget to provide guidance to federal agencies in initiating, developing, using, and maintaining enterprise architecture. The FEAF is based on the Zachman architecture and consists of the following layers: planner perspective, owner perspective, designer perspective, builder's perspective, and subcontractor's perspective. Each of these layers is associated with the columns data architecture, application architecture, and technology architecture. The FEAF matrix is depicted in Fig. 23. The "x" in Fig. 23 indicates that the column dimension should be addressed.

	Data Architecture	Application Architecture	Technology Architecture
Planner Perspective	x	x	x
Owner Perspective	x	x	x
Designer Perspective	x	x	x
Builder's Perspective	x	x	x
Subcontractor's Perspective	x	x	x

Fig. 23. Federal enterprise architecture framework.

2) Composite Process-Personalization Semantics

The six types of CPP semantics identified in this book are the knowledge semantics (KS), rules semantics (RUS), roles semantics (ROS), users-profile semantics (UPS), infrastructure semantics (INS), and communication semantics (COS). The role of these semantics in composite services development with CPP is defined in Chapter IV.

KS provides a representation of the knowledge within an enterprise. According to knowledge-based views of the firm, knowledge is the most important strategic resource of the firm, and the firm adds value to its businesses by providing superior organizing principles for creation, transfer, integration, and leverage of knowledge [252]-[256]. This knowledge is present in resources such as manuals, letters, summaries of responses to clients, news enterprise knowledge, customer information, competitor intelligence, and knowledge derived from work processes. Knowledge-based resources refer to the ways in which the more tangible input resources are manipulated and transformed so as to add value [257], [258]. The three notable properties of knowledge are tactiness, context specificity, and dispersion. Tactiness is the extent to which knowledge is or is not codifiable [259], [260]. Context specificity is the extent to which knowledge is highly contextualized and co-dependent on unidentified aspects of the local environment [261]. Dispersion is the extent to which knowledge is concentrated in the head of an individual or spread out across the minds of many [262]. A composite service that takes into account the KS of the enterprise can be developed to aid and assist the user in a more informed and interactive manner.

RUS describes the business rules that provide a representation of the conditions that must be satisfied, and their role is to determine how operational decisions within an organization must be made [263]. Morgan describes a rule as a constraint. It defines the condition under which a process is carried out or the new conditions that exist after a process is completed [264].

ROS describes roles that provide a representation of the set of users that are responsible for a task. Roles also define the set of permissions to provide access control [265]. According to [34], roles provide the following definitions:

- Users associated with a role.
- Resources private to roles required to participate in the process.
- Activities carried out by roles to manipulate resources.
- Interactions between roles to transfer resources.
- States of a process in terms of logical conditions that control the execution and validation of activities.

UPS provides a representation of the users that are participating in a process and their preferences. UPS is associated with the roles of a task and is comprised of such information as the user's identity, their specialty, qualification, location, and preferences. INS describes the type of resources that the user requires for completing a task. INS is associated with UPS and describes the type of resources that the user requires for completing a task. COS represents the input and output messages of the different tasks in a process.

K. Other Issues - Integration of Semantics

This section is intended to express that several approaches can be used to represent the different semantics in machine-accessible form. Several approaches exist to make KS machine-accessible by composite services systems, such as knowledge management techniques to make knowledge accessible and reusable to the enterprise in a machine-accessible form [266]. Several studies recommend steps to identify, organize, and represent knowledge in a machine-accessible form [17], [228], [267], [268].

RUS can be represented in machine-accessible form using several approaches [198], [215]. For instance, the rule markup initiative (RuleML) allows the representation of rules in XML [212]. Another popular approach is the use of expert systems, such as JESS [74], [269], an expert system shell and scripting language written entirely in Java. JESS has a syntax based on C Language Integrated Production System (CLIPS) to describe rules [66]. Appendix C shows a simple rule system for book recommendation.

ROS-based access control can be provided in several ways [270], [271]. Ferraiolo and Kuhn are credited as the first to formalize role-based access control [272]. NIST describes a standard model for role-based access control [273].

Ontology can be developed to store the UPS. For example, the friend of a friend (FOAF) project is an ontology that describes people, the links between them, and the things they create and perform [274]. Jian and Tan describe an ontology model for personalized Web services [275]. An initial description of the user can be added to the composite services system, and the system can then enhance the user description using such techniques as learning agents [276].

There are several approaches for describing INS. For instance, Web service and semantic Web standards discussed in Chapter II can be used to describe the services [9], [41]. A popular approach for representing COS is using XML-Schemas [277]. RDF and OWL can also be used for representing COS [40].

L. The Task System Model

The description of the task system model in this section is based on [29]-[31]. The task system model provides a powerful approach for process modeling and analysis. The task system model is used as an abstract process modeling approach in this book. The task system model represents the activities of a process or process system as a set of tasks

$$\tau = \{T_1, T_2, ..., T_N\}. \tag{3}$$

The tasks of a task system model are not interpreted. That is, beyond the association of tasks with computational activity, there are no assumptions made regarding the semantics, granularity, or functionality of a task or task system. Therefore, a single task may represent anything from a number of steps in the design of a software artifact to a single computer instruction that makes the task system appropriate to model different process layers. Associated with each task is an initiation event denoted as $T_i \uparrow$ and a termination event, denoted as $T_i \downarrow$.

In general, any process comprising multiple tasks requires synchronization among at least a subset of the tasks in order to ensure correct execution. Synchronization requirements determine the necessary execution order of tasks relative to one another. Assuming synchronization to be an inherent attribute of the process representation scheme, a task system is defined as the pair

$$C = (\tau < *). \tag{4}$$

In (4), the precedence relation $< *$ describes the ordering of pairs of tasks from the set τ.

1) Precedence Relations and Precedence Graph in a Task System

A graph $G = (V, E)$ is described and analyzed in terms of its vertices V and its edges E. Directed graphs have a direction associated with each edge, generally indicated by an arrowhead. A directed, acyclic graph (DAG) is a directed graph having the characteristic that no traversal of the graph can visit any vertex more than once. The requirement that no cycles exist greatly simplifies many graph algorithms but may limit the attached semantics, depending upon the application.

A directed, acyclic graph can be interpreted as indicating an order or precedence among the vertices. Within the field of computer science, this interpretation has been widely used in such application areas as operating systems theory and compiler design [28], [278].

In the context of the task system model, a directed acyclic graph $G = (V, E)$ is well suited for representing the precedence relation $< *$ among a set of tasks τ, where the set of vertices V of the graph are interpreted to represent the set of tasks τ and the set of edges E are interpreted to represent precedence relations $< *$ for some task system $C = (\tau < *)$. The anti-reflexive and anti-symmetric properties of the precedence relation are enforced by the acyclic nature of the corresponding graph. The transitive property of the precedence relation is expressed through reachability within the corresponding graph; that is, the ability to reach a vertex v_j from another vertex v_i, where $v_i, v_j \in V$ by traversing edges of the graph.

A directed, acyclic graph $G = (V, E)$ that represents a task system $C = (\tau < *)$ is referred to as a precedence graph. The notion of precedence graphs is formalized in the following definition: The precedence graph $G_C = (V, E)$ corresponding to a task system $C = (\tau < *)$ has the following characteristics [28]:

- A vertex $v_i \in V$ if and only if there is a corresponding task $T_i \in \tau$.
- An edge $(v_i, v_j) \in V$ if and only if $T_i < *T_j$ and there exists no T_k such that $T_i < *T_k < *T_j$.

The second stipulation ensures that no redundant specifications of precedence appear in the precedence graph.

2) Tasks in a Task System

Tasks in the task system model represent real-world activities associated with a process. An initial task is a task with no predecessors, while a task with no successors is a terminal task. An initial decision point task is a task having multiple successors, while a task having multiple predecessors is a terminal decision point task. An intermediate task is a task that has a single predecessor and a single successor.

A closed task system is a system that has a single initial task and a single terminal task. A task system $C = (\tau < *)$ that is not closed can be transformed to a closed task system through the following steps: First, by adding a "dummy" initial task T and a "dummy" terminal task T' to τ, and second, by adding to C the relations $T < *T_i$ for each initial task $T_i \in \tau$ and the relations $T_j < *T'$ for each terminal task $T_i \in \tau$. Closed task systems simplify concatenation and tailoring operations [30].

3) Task System Chains

Two tasks systems T and T' are sequential if $T <*T'$. For a task system $C = (\tau < *)$, a set of sequential tasks $\tau' \subseteq \tau$ and the associated precedence relations $C' \subseteq C$ that contains an initial task T and a terminal task T' is referred to as a chain. A task system may contain any number of chains, and any number of chains may exist between a pair of tasks T and $T' \in \tau$. A unique chain $C^x = (\tau^x < *)$ is distinguished by the fact that for each task $T \in \tau^x$, T is neither an initial decision point nor a terminal decision point task.

We can view any chain as a task system, and any task system can be composed of one or more chains. A single task T forms a trivial chain, with T serving as both the initial and terminal task for the chain, so that $\tau' \equiv \{T\}$ and $C' \equiv \Phi$.

4) Resource Types

Resource types define the task inputs and outputs represented in the task system. We can represent a resource type as $RT_x \equiv \{z_1, ..., z_n\}$, where n is the number of attributes required to sufficiently describe a resource of type X. The attribute z_i associated with a resource type comprises the characteristics of the corresponding resources that are observed or modified by the tasks of a task system C. The nature of the resources of the task system model is also un-interpreted, indicating that it can represent a single resource or a multitude of resources.

5) Finite State Model of a Task System

Delcambre introduced a finite state automation (FSA) model for task system execution modeling [31], [32]. Fig. 24 depicts the FSA of the task system model. Formally, a FSA is a 5-tuple $(Q, \Sigma, \partial, q0, F)$ where Q is a finite set of states, Σ is a finite alphabet of input symbols, ∂ is a transition function with $\partial : Q \times \Sigma \rightarrow Q$, $q0 \in Q$ is an initial state, and $F \subseteq Q$ is a set of final states.

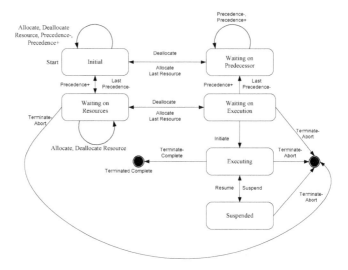

Note: From "A Resource-Focused Framework for Process Engineering," by S.F. Mills, 1995, Ph.D. Dissertation, Dept. School of Engineering and Applied Science, Southern Methodist University. Copyright 1997 by S.F. Mills. Reprinted with permission.

Fig. 24. Finite state model of the task system.

A task T can assume the following set of states Q_E during execution:

- In the dormant state, the task T is waiting on both the completion of predecessor task and the allocation of resources. This is the initial state of all tasks.

- In the waiting on predecessor state, the task T is waiting on the termination of one or more predecessor tasks.

- In the waiting on resources state, the task T is waiting on the allocation of one or more resources.

- In the waiting on execution state, the task T is waiting for initiation in order to begin execution.

- In the execution state, the initiation event $T\uparrow$ has occurred, and task T is executing.

- In the terminated abort state, the task T has been aborted, and the aborted termination event $!T \downarrow$ has occurred.
- In the termination complete state, the termination event $T \downarrow$ has occurred, and task T has terminated normally.
- In the suspended state, the execution of task T has been suspended. The reasons for suspensions are not interpreted in the task system model.

M. Interaction-Pattern-Based Process Mining

In this section, an interaction-pattern-based process mining approach is described for process analysis and optimization is discussed. Process mining for analysis and optimization of composite services can be performed after composition or during the process modeling stage. In the latter case, process enactment would be used to simulate the data. Delcambre observes that the term process enactment is synonymous with process execution [31]. She states that process enactment is important in process engineering as it may provide for the testing of new descriptions, use in an actual limited-scope development, or execution in an actual environment.

Process mining and analysis is performed because a process can deviate from its process model at run time because of several reasons [279]. It provides an approach to analyze the performance of a process model at modeling or run time. The results of the analysis can be used to optimize the process model. Deviations in a process usually occur because of the human element [280]. The deviations in a process can occur because of the following reasons [281]:

- The process definitions omit or do not allow for relevant project contingencies.
- Sometimes risks are taken.
- Some process definitions are more amenable to deviations.
- People have good ideas, some of which are better than the defined processes.
- The process does not make sense either because of individual differences or because of lack of training.
- A lack of commitment or interest.

Deviations in process can also occur during the development of composite services; e.g., if a service needed for the development of a composite service is not available for some reason. The composite service might have to be realized with another service or a set of services that perform the given task, which can result in deviations in the run-time process.

Although process analysis is considered an important aspect of current process tools, they fall short of the objective [279]. The reason is that these tools only focus on measuring performance indicators such as flow time and utilization and are not effective in discovering the process and its organizational context.

Mining and analyzing interaction patterns are important for CPP. Alexander describes an interaction pattern as a three-part rule, which expresses a relation between a certain context, a problem, and a solution [230]. Dustdar and Hoffman observe that discovering complex interaction patterns offers additional knowledge about the role of actors within an organization [279]. In particular, interaction pattern detection is important in the case of highly dynamic processes, such as processes involving human interactions.

An interaction pattern is a representation of the process of the system. It is based on the communications among the resources of a system. There are several examples of interaction patterns in literature [66], [212], [279], [282]-[284]. An example of a simple interaction pattern is the request-reply interaction pattern in which a sender sends a request and waits for a response [285]. The response can be synchronous or asynchronous.

A complex interaction pattern is the 4-eyes principle [212]. The 4-eyes principle is also referred to as a separation of duties. The 4-eyes principle is common in a collaborative process, such as when two or more people make decisions independent of one another. In extreme cases, the users may be unaware of the identity of the each other. In other cases, the other users just offer additional opinion.

1) Interaction Pattern Mining Using Conant's Decomposition Approach

Process mining is an effective method of identifying and modeling interaction patterns [158]. Most process mining approaches are based on mining transaction logs, such as audit trails or workflow logs [279]. Dustdar and Hoffman propose a method to mine interaction patterns using Social Network Analysis. SNA is a method that focuses on the analysis of relationships among social entities, and on the patterns and implications of these relationships [286]. They describe SNA rules and procedures for pattern analysis.

A model of directed and connected graphs for process mining is employed in [224]. They consider interaction patterns as a model that can represent the combination of simultaneous events. An event is an element of interaction that has a start time and an end time [224]. An episode is defined as a set of events. The model in [224] considers only simultaneous patterns for modeling and ignores temporal information.

The process mining approach developed in this book is based on Conant's decomposition approach, which is a technique to decompose a system based on its subsystems interaction [62], [63]. Conant proposes a measure of the intensity of interactions between subsystems. The intensity measure is the normalized transmission t_{ij}.

Conant's approach is based on Simon's postulates of complex system that are as follows [1]: The short-run behavior of each of the component subsystems is approximately independent of the short-run behavior of the other components. As a corollary, the short-run behavior of each of the parts within a subsystem is not approximately independent of all other parts in its subsystem. Therefore, a measure of intensity can be used to decompose a system and mine its interaction patterns. A detailed discourse of Conant's method is provided in [19], [62], [63].

The steps to calculate the intensity measure t_{ij} is summarized as follows [287]:

1) Consider a complex system that has K primary variables, each of which is observed once every standard time increment N, giving a total of $K \cdot N$ observations. With each of the primary variables is associated a derived variable $X_j, 1 \leq j \leq K$ that can be grouped into sets.

2) If the K primary variables are not integer metrics then derived integer variables, X_j, $1 <= j <= K$, are chosen to represent the subsystem categories.

3) The variables are grouped into sets of time series observations.

4) The number of occurrences, n, of each possible value in each set is counted and used to calculate the entropy of that set by

$$H\left(X_j\right) = log_2 N - \frac{1}{N}\sum_{i=1}^{M_j} n_i \, log_2 \, n_i \qquad (5)$$

where j is the jth variable, $H\left(X_j\right)$ represents the entropy of the variable X_j, N is the total number of observations, M_j is the upper limit of the range of values of X_j, i = index value, and n_i = number of observations of each index value.

5) Vectors representing pair-wise combinations of each set of a variable's time series observations with the second variable offset by one time step are generated.

6) The number of occurrences of each possible combination of values in each vector is counted for each vector and used to calculate the joint entropy $H\left(X_i, X_j'\right)$, where X_j' represents the set of observations offset by one time increment.

7) The transmission parameter is then calculated by
$$T\left(X_i : X_j'\right) = H\left(X_i\right) + H\left(X_j'\right) - H\left(X_i, X_j'\right). \qquad (6)$$

8) The normalized transmission t_{ij} can be obtained by dividing $T(X_i : X'_j)$ values by $H(X'_j)$
$$t_{ij} = \frac{T(X_i : X'_j)}{H(X'_j)} \, . \qquad (7)$$

Appendix B provides the code and screen shots of the implemented Conant-Web-based application. The steps to calculate t_{ij} are described with an example. Consider a composite service with three tasks X_1, X_2, and X_3. Each of the tasks has two states that are empirically observed as 1 and 2. A random generator is used to generate a set of 10 primary variables indicating 10 observations with some standard time increment. Table 14 shows the set of input variables. Next, the variables are grouped into sets of time series observations to calculate $H\left(X_i\right)$ and $H\left(X_j'\right)$. The first nine observations of Table 14 are used for calculating $H\left(X_i\right)$ and the variables from the second to tenth observations for calculating $H\left(X_j'\right)$. Therefore, N is nine in this case. Since the number of tasks is three, j equals three. Table 15 represents the grouping of the variable into sets of time series observations for calculating $H\left(X_i\right)$.

Table 16 represents the grouping of the variable into sets of time series observations for calculating $H\left(X_i'\right)$.

TABLE 14
SET OF INPUT VARIABLES

j\i	1	2	3	4	5	6	7	8	9	10
1	1	1	2	1	2	2	2	2	1	2
2	2	1	2	2	1	1	2	1	2	1
3	1	1	2	1	2	2	1	1	1	2

TABLE 15
GROUPING OF THE INPUT VARIABLES INTO SETS OF
TIME SERIES OBSERVATIONS FOR $H\left(X_i\right)$

nj\i	1	2	N
1	4	5	9
2	4	5	9
3	6	3	9

$H\left(X_1\right)$ is calculated from (5) as $log_2 9 - \dfrac{1}{9}\left(4*log_2 4 + 5*log_2 5\right)$ to give 0.9910. Similarly, $H\left(X_2\right)$ and $H\left(X_3\right)$ can be calculated. The values are 0.9910 and 0.9182, respectively. The data in Table 16 is used to calculate the values of $H\left(X_1'\right), H\left(X_2'\right)$, and $H\left(X_2'\right)$. The computed values are 0.9182, 0.9910, and 0.9910, respectively.

TABLE 16
GROUPING OF THE INPUT VARIABLES INTO SETS OF
TIME SERIES OBSERVATIONS FOR $H\left(X_i'\right)$

Nj\i'	1	2	N
1	3	6	9
2	5	4	9
3	5	4	9

After the above step, a table of vectors is constructed by concatenating pairs of values from Table 14 to calculate the joint entropy $H\left(X_i, X_j^{'}\right)$. The value in the first column represents the time interval. Values in subsequent columns represent combinations of the values from the columns in Table 14, with the second value offset by one time increment. For example, to construct the value for time interval 1, column $1,1^{'}$, the values from time interval 1, column 1, Table 14 are concatenated with the values from time interval 1, column 2, Table 14. This process yields the value of $1,1^{'}$. Similarly, the other values are generated and displayed in Table 17.

The entropy of each column vector is calculated using

$$H\left(X_a X_b^{'}\right) = log_2 N - \frac{1}{N}\sum_{i=1}^{M_j} n_i \, log_2 \, n_i \tag{8}$$

where subscripts a and b are used to avoid confusion with the summation indices, $j = j^{th}$ vector, $H\left(X_a X_b^{'}\right)$ = entropy of the vector $X_a X_b^{'}$, N = total number of observations, M_j = number of values of $X_a X_b^{'}$, i = index value, n_i = number of observations of each index value. For this data set, there were three tasks, which result in nine combinations. Therefore, j equals nine. The vectors $X_a X_b^{'}$ were the nine sets of data from Table 18. N equals nine and M_j equals four because there are two states.

TABLE 17
VARIABLES TO CALCULATE THE JOINT ENTROPY $H\left(X_i, X_j^{'}\right)$

$i,j^{'}$	1	2	3	4	5	6	7	8	9
$1,1^{'}$	1,1	1,2	2,1	1,2	2,2	2,2	2,2	2,1	1,2
$1,2^{'}$	1,1	1,2	2,2	1,1	2,1	2,2	2,1	2,2	1,1
$1,3^{'}$	1,1	1,2	2,1	1,2	2,2	2,1	2,1	2,1	1,2
$2,1^{'}$	2,1	1,2	2,1	2,2	1,2	1,2	2,2	1,1	2,2
$2,2^{'}$	2,1	1,2	2,2	2,1	1,1	1,2	2,1	1,2	2,1
$2,3^{'}$	2,1	1,2	2,1	2,2	1,2	1,1	2,1	1,1	2,2
$3,1^{'}$	1,1	1,2	2,1	1,2	2,2	2,2	1,2	1,1	1,2
$3,2^{'}$	1,1	1,2	2,2	1,1	2,1	2,2	1,1	1,2	1,1
$3,3^{'}$	1,1	1,2	2,1	1,2	2,2	2,1	1,1	1,1	1,2

TABLE 18
GROUPING OF THE INPUT VARIABLES INTO SETS OF
TIME SERIES OBSERVATIONS FOR $H\left(X_i, X_j'\right)$

i,j'	1,1	1,2	2,1	2,2	N
1,2'	3	1	2	3	9
1,3'	1	3	4	1	9
2,1'	1	3	2	3	9
2,2'	1	3	4	1	9
2,3'	2	2	3	2	9
3,1'	2	4	1	2	9
3,2'	4	2	1	2	9
3,3'	3	3	2	1	9

The entropy of vector 1,1' was calculated as

$$H(1,1') = log_2 9 - \frac{1}{9}(1*log_2 1 + 3*log_2 3 + 2*log_2 2 + 3*log_2 3) \tag{9}$$

$$H(1,1') = 3.169 - \frac{1}{9}(0 + 3*1.589 + 2*1 + 3*1.589) \tag{10}$$

$$H(1,1') = 3.169 - 1.2815 = 1.8875. \tag{11}$$

Similarly, the values of $H(1,2'), H(1,3'), H(2,1'), H(2,2'), H(2,3'), H(3,1'),$ $H(3,2'),$ and $H(3,3')$ were calculated to be 1.8910, 1.7527, 1.8910, 1.7527, 1.9749, 1.8365, 1.8365, and 1.8910. The transmission metrics can be calculated from the entropy values calculated in steps 2 and 4 using (6). The transmission metric for trunk group 1 to itself one time interval later was calculated as

$$T(1:1') = H(1) + H(1') - H(1,1') \tag{12}$$

$$T(1,1') = 0.9910761 + 0.9182957 - 1.8875 \tag{13}$$

$$T(1,1') = 0.0217. \tag{14}$$

The transmission metrics for the nine permutations of trunk group combinations are shown in Table 19. The normalized transmission t_{ij} was calculated using (7)

$$t_{11} = \frac{0.0217}{0.9182} \qquad (15)$$

$$t_{11} = 0.0236. \qquad (16)$$

TABLE 19
CALCULATED t_{ij} VALUES

$I j'$	$X1'$	$X2'$	$X3'$
$X1$	0.0217	0.0919	0.2315
$X2$	0.0199	0.2315	0.00727
$X3$	0	0.0734	0.0184

N. Conclusion

In this chapter, an overview of process engineering of composite services was provided. Since the term "process" has several different connotations, a definition was provided to clarify its use in this book. The process life cycle that forms an integral part of the composite $P^2_{\text{FRAMEWORK}}$ was described. The different process technologies and process formalisms that can be used in process engineering was described. The process-oriented composite services development approaches were classified.

An overview of enterprise architecture frameworks was also provided. The six types of semantics identified in the CPP dimensions and their integration were discussed. The task system model that is used to model the processes of the case study examples was discussed. The process mining approach for analysis of composite services was discussed.

IV. THE COMPOSITE, INTEGRATION, AND PERSONALIZATION ARCHITECTURE FRAMEWORK FOR COMPOSITE SERVICES

In this chapter, the dimensions required to address the CPP challenge are defined. The composite $P^2_{FRAMEWORK}$, an architecture framework for guiding composite services development with emphasis on CPP is described. The service-agent model is also described. Two case studies that demonstrate the composite $P^2_{FRAMEWORK}$ guidance are provided. A CPP development model is described. The interaction-pattern-based process mining approach is demonstrated.

A. Overview of the Book Contribution

Consider the steps in process-oriented composite services development. In the first step, process models that consist of both the abstract and executable process models are developed. Abstract process modeling starts with a high-level model of the enterprise with minimum details. Models with increasing levels of semantics follow this high-level model. Each stage of the abstract process modeling can be associated with process analysis to improve the model. Once a model of sufficient detail is realized, the model is transformed to an executable process model such as BPEL to develop composite services.

The problem of supporting CPP occurs at two stages because of loss of semantics of user needs. First, the loss of semantics occurs at the abstract process modeling stage because the semantics captured by abstract process models might not be as comprehensive as needed. Second, there is loss of semantics during the transformation from abstract process models to executable process models. This is because executable process modeling technologies are inadequate in their support to configure and manage the captured semantics. Harrison-Broninski describes this limitation in his book [34], showing that executable process technologies were designed to help technicians build automated process execution engines capable of orchestrating distributed computing resources of various kinds. Harrison-Broninski further observes that the current work on executable process technologies is driving it downwards towards programming rather than upwards, which makes it more difficult to configure and manage semantics. It is important that the executable process technologies be driven downwards towards programming to support the efficient composition of services. It follows that there must be a mechanism to support the executable processes with semantics to develop composite services with CPP. Fig. 25 depicts the stages at which the loss of semantics can occur in process-oriented composite services development.

A systematic framework that guides the development of composite services with an emphasis on CPP is required to address the loss of semantics. The core contribution of this book is a composition, integration, and personalization framework, the composite $P^2_{FRAMEWORK}$ that provides systematic guidance for composite services development with CPP.

The composite $P^2_{FRAMEWORK}$ leverages a process-oriented approach for composite services development. The guidance the framework offers is based on the three dimensions of CPP. The composite $P^2_{FRAMEWORK}$ uses the service-agent model that combines the services concept with the agent concept for configuring and managing the semantics of user needs in composite services. A concept map is developed that visualizes and articulates the relationship of the CPP semantics with the service-agents of a task. The relationships articulated by the concept map can be used to model and develop composite services with CPP.

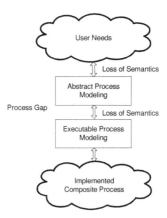

Fig. 25. Loss of semantics in process-oriented composite services.

Two case studies are used to demonstrate and validate the guidance offered by the composite P^2FRAMEWORK for developing composite services with CPP. The first case study is the WCS process with a human interaction task. The second case study is the GLIP part of the composite epidemiology research services process for identifying candidate genes for obesity research. The GLIP case study involves complex interactions between the different participants in the research process.

The interaction-pattern-based process mining approach is also demonstrated using the WCS case study. The process mining can be used in combination with the CPP semantics for process analysis and optimization.

B. Composite Process-Personalization Dimensions

A CPP framework must offer guidance along three dimensions. The primary dimension is the semantic-dimension. The secondary dimension is the syntactic-dimension. Fig. 26 depicts the CPP dimensions.

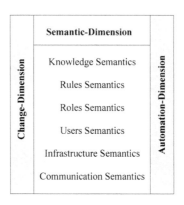

Fig. 26. Dimensions of composite process-personalization.

Six types of semantics are identified in the semantic-dimension: KS, RUS, ROS, UPS, INS, and COS. Their role in providing CPP is defined as follows:

- KS answers specific questions required to address a task.
- RUS provides specific conditions required to perform a task.
- ROS provides a representation of the group of users interacting with a task and their access permissions.
- UPS provides a representation of the users associated with a task and their preferences.
- INS provides a representation of the mechanistic resources and tools that are needed to perform a task.
- COS provides a representation of the input and output messages that are involved in performing a task.

Examples of the six types of semantics are provided in the composite epidemiology research services process case study. The change-dimension and the automation-dimension are the two types of syntactic-dimension. Enterprise processes, especially those with human interactions constantly change. As people discover new information or begin to understand existing processes, they tend to modify the process to accommodate their new knowledge [34]. Composite services with CPP should be flexible to accommodate this need for change. Another aspect to the change-dimension is process analysis. The results of the analysis can then be used to optimize the process, which can also improve the interactions of the users of the system. Optimizing can be achieved either by organizing or adding more resources to the process, or by automating the process. The automation-dimension is critical because it helps to reduce the workload of the user by automation of a process. The automation-dimension also improves the efficiency of the process, as it reduces the need for human interaction. The semantic-dimension supports the syntactic-dimension. Besides enhancing the interactions of users, the semantic-dimension also aids in both process analysis and the enhanced composition of services.

C. A Service-Agent Model for Composite Process-Personalization

The service-agent is an abstraction that provides a modeling approach for the development of composite services with CPP. It also provides a "transformation" link between the abstract process models and the executable process models. It combines two concepts, services, and agents. The services concept provides the abstraction for the modeling of different types of resources. The agents concept provides support for configuring and managing semantics to address the CPP needs of users. The service-agent can be used to represent mechanistic and people resources. The concept map, depicted in Fig. 27, shows the linking of the tasks to the resources using the service-agent.

The concept map depicted in Fig. 27 can be described as follows: A task can have multiple service-agents. The service-agents can be of two types, mechanistic or people. Mechanistic service-agents are abstractions of composable resources, such as services, methods, and programs. People service-agents are abstractions of human activity. Every task requires input messages for operation and generates output messages. The input and output messages are types of INS. The input messages are required or provided by service-agents. For instance, input messages are required to invoke mechanistic resources, such as Web services. Input messages can also be provided by a service-agent to trigger a task. A service-agent may require knowledge for performing a task. Knowledge answers the specific needs of service-agents to perform a task. Rules specify the conditions necessary to carry out a task.

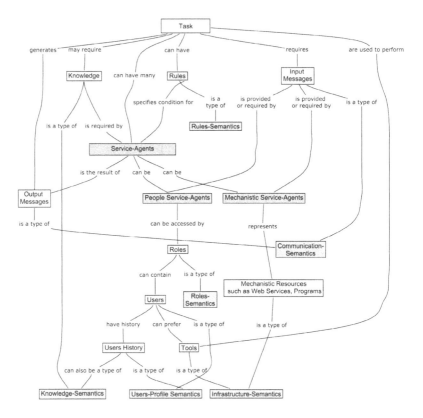

Fig. 27. Concept map for task representation for CPP.

The concept map also shows that a people service-agent can be accessed by certain roles of users that are specified by ROS. Each role can have a set of users specified by UPS. Users can use certain tools, a type of INS, to perform their tasks. Users can also have history, a type of KS, which can be used to tailor their interaction with the process. The representation of the tasks of a process using the relationships described in the concept map is demonstrated in the case study examples.

The service-agent model also supports the syntactic-dimension of CPP through facilitating enhanced composition and analysis of composite services. First, using agents allows an analyst to enact the process and simulate the human interactions. The simulations can then be used to analyze and optimize the process. Second, it provides support for the enhanced composition of services. The service-agent achieves this by providing a description of the composite process and its resources enhanced with the CPP semantics. Therefore, the composition engine can use the semantics for a more precise discovery, selection, and composition of services that match the user needs.

The service-agent provides support for mapping to the syntax of current executable process-models. The specific mapping would depend on the executable process modeling approach. A potential mapping of the service-agents to BPEL [9], an executable process language, is discussed. In BPEL, mechanistic resources, such as services, methods, and programs are composed as partner links. Partner link is an abstraction of mechanistic services used to describe how two parties interact and what each party offers [9]. Therefore, the mechanistic service-agents can be mapped to partner links. Support for the integration of human activity is not inherent in BPEL, although the new proposed specification of BPEL4People aims to address this limitation. Consequently, the support for integration of human activity is left to the developers of the composite services system and their chosen tools. Typically, developers have used ad hoc mechanisms for integrating human activity. For example, a popular mechanism for integrating human activity is by considering the activity as an asynchronous service, and then representing the asynchronous service as partner links. The client that is built to interact with the BPEL process provides support for human interaction. A similar approach can be used to map the people service-agent but the support for human activity comes from the semantics added as properties of agents.

The advantage of the service-agent approach over ad hoc approaches for integrating human activity is that the semantics are systematically derived from the abstract process models and mapped to the executable process models. As in the ad hoc approaches used currently, the human interactions are integrated as services. However, several disadvantages of the ad hoc approaches are overcome by the service-agent approach. First, the integration of the semantics of user needs is performed in a systematic manner. In the ad hoc approaches, the lack of a systematic approach implies that there could be a loss of semantics at the executable process modeling stage. Second, in the ad hoc approaches, instead of being incorporated in the processes, the semantics of user needs are incorporated in the client of the system. That is, the semantics are integrated outside the scope of the executable processes. Therefore, the process analyst would have to rely on documentation or other mechanisms to factor the semantics in his analysis. If the analyst is not careful or if the documentation is insufficient, the results of the process analysis would be incomplete. The service-agent overcomes this second limitation by incorporating the semantics as properties of agents. The analyst can incorporate the semantics in his analysis by simulating the agents.

D. The Composite $P^2_{FRAMEWORK}$ for Composite Process-Personalization

The composite $P^2_{FRAMEWORK}$ provides systematic guidance for composite services development with CPP. Three layers are defined in the composite $P^2_{FRAMEWORK}$: core competency layer, business model layer, and composite services layer. The core competency and business model layers are defined to provide a direction and scope for the integration of large systems. The primary focus of our work is on the composite service layer. It builds on a process-oriented approach of composite services development and uses the service-agent model for incorporating CPP in composite services.

1) Core Competency Layer

The core competency is the strength of the enterprise [288]. It is derived from the resources of the enterprise, such as its rules, policies, and knowledge of its employees. Therefore, the process of identifying the core competency implies that the enterprise can identify the scope of the integration in a systematic and focused manner. Hamel and Prahalad first coined the term core competency [289]. Core competency is described as the collective learning in the organization, especially the capacity to coordinate diverse production skills and integrate streams of technologies. The importance of core competency is underscored by citing the examples of Japanese firms in the 1980s. Japanese firms such as Canon, Honda, and NEC gained a competitive advantage by effectively leveraging its core competencies. These firms were able to obtain a creative advantage over the resource-rich enterprises in the United States and Europe. Another example of the importance of leveraging core competencies, showing that Sony and Dell are successful because they were able to leverage their core competency across different markets is provided in [47]. Core competency is defined as a combination of complementary skills and knowledge bases embedded in a group or team that result in the ability to execute one or more critical processes to a world-class standard [290]. A core competency should be applicable to a wide variety of markets, make a significant contribution to the benefits of the product as perceived by the customer, and be difficult for competitors to imitate [289].

2) Business Model Layer

A business model is a conceptual tool that contains a large set of elements and their relationships and allows the expression of the business logic of a specific firm [291]. It is a description of the value a company offers to one or several segments of customers and of the architecture of the firm and its network of partners for creating, marketing, and delivering this value and relationship capital to generate profitable, and sustainable revenue streams. A business model answers the following questions [292]: who are the value-adding business actors involved; what are the offerings of which actors to which other actors; what are the elements of offerings; what value-creating or value-adding activities are producing and consuming these offerings; which value-creating or value-adding activities are performed by which actors. Os-terwalder and Pigneur suggest that a business model provides a good conceptual and architectural foundation for the development of processes [293].

3) Composite Services Layer

The primary focus of this book is the composite services layer. The composite services layer is based on a process-oriented development of composite services. The life cycle of the composite services layer consists of process modeling, process composition, and process analysis and optimization. The implementation steps are based on the EM, as discussed in Chapter III; that is, the process modeling stage starts with the development of abstract process models. The abstract process models are then transformed to executable process models for composition based on the mapping provided by service-agents. There is constant feedback within and between the layers to ensure that validated and verified (V&V) systems are developed. For instance, the task system model provides an approach for identifying deadlocks, synchronization, and cohesion problems [28]. We also describe a process mining approach for analysis of composite services in Section H.

Fig. 28 depicts the composite $P^2_{FRAMEWORK}$ matrix. It describes the composite $P^2_{FRAMEWORK}$ as a process-engineering framework with three layers: core competency layer, business model layer, and composite services layer. Each of these layers has six associated columns based on the semantic-dimension. KS answers specific questions required to address a task. RUS provides specific conditions required to perform a task. ROS provides a representation of the group of users interacting with a task and their access permissions. UPS provides a representation of the users associated with a task and their preferences. INS provides a representation of the mechanistic resources and tools that are needed to perform a task. COS provides a representation of the input and output messages that are involved in performing a task.

"Primary" and "secondary" notations are used in the columns, as depicted in Fig. 28, to indicate that the composite services layer is the primary focus of my book. The other layers are secondary focus. The matrix provides systematic guidance for capturing semantics during the integration of large systems with composite services. In order to configure and manage semantics, the service-agent model is used.

Layers/ Columns	Semantic-Dimension					
	Knowl-edge	Rules	Roles	Users-Profile	Infrastructure	Communication
Core Compe-tency	Secondary	Secondary	Secondary	Secondary	Secondary	Secondary
Business Model	Secondary	Secondary	Secondary	Secondary	Secondary	Secondary
Composite Services	Primary	Primary	Primary	Primary	Primary	Primary

Fig. 28. Composite $P^2_{FRAMEWORK}$ matrix for CPP.

The composite $P^2_{FRAMEWORK}$ leverages process-oriented composite services development approach. Processes will guide the composition of services. There are several advantages of this approach. Composite services provide standards and powerful tools for composition of services [43-45]. Second, we know that enterprises are organized as processes [9], [32], [34], [37]. This implies that a process-oriented approach provides a natural and systematic approach for modeling and developing com-

93

posite services. The rich set of process engineering modeling technologies and formalisms can also be leveraged with the guidance offered by the framework to develop composite services with CPP. A process-oriented approach also implies that we can use process enactment to analyze process models, reducing the errors in composite services development. The service-agent model can be used for incorporating CPP in process-oriented composite services development. This implies that the large systems developed based on the composite $P^2_{\text{FRAMEWORK}}$ can systematically incorporate CPP, providing an advantage over those implemented with other frameworks. The guidance offered by the composite $P^2_{\text{FRAMEWORK}}$ to develop composite services with CPP leveraging process-oriented composite services development is demonstrated using two case study examples.

E. Case Study – Weather Composite Services

The first case study is an extended version of the WCS example discussed in Chapter II. The example includes an approval task, a human activity, to indicate whether the WCS should email users with the weather report. The extended version of the WCS is depicted in Fig. 29. The example is used to show the development steps of composite services with CPP, starting from abstract process modeling to executable process modeling, based on the guidance of the composite $P^2_{\text{FRAMEWORK}}$.

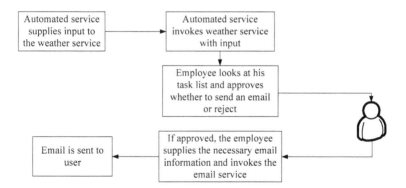

Fig. 29. Extended WCS example.

The task system model is used to represent the abstract process model of the WCS. Fig. 30 depicts the task system model of the WCS. The WCS task system can be represented as $C_{WCS} = (\tau, <*)$, where τ is the set of tasks and $<*$ is the precedence relation on τ. The set of tasks is

$$\tau_{WCS} = \{\text{Weather, Approval, Email, Reject}\}. \tag{17}$$

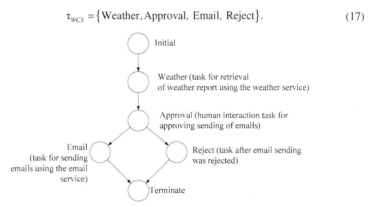

Fig. 30. Task system model of the WCS.

The weather task involves invoking the weather service, supplying the necessary parameters, and receiving the report. The approval task represents the process of an employee looking at the weather report and verifying whether to send an email to another user or reject the email. The email task represents the process of invoking the email service, supplying the necessary information, and sending the email. The reject task represents the process of choosing not to send an email to the user. For the purpose of this example, the reject task is considered as an empty task with no resources.

The task system $C_{WCS} = (\tau, <*)$ can be defined as

$$C_{WCS} = \{(\text{Weather, Approval}), (\text{Approval, Email}), (\text{Approval, Reject})\}. \tag{18}$$

The resource types of the task system model can be defined as

$$\begin{aligned}
\rho_{WCS} \equiv \{\ &RT_{weatherservice}, RT_{automatedservice}, RT_{weatherparameters}, RT_{weatherreport}, \\
&RT_{employeeknowledge}, RT_{employeerole}, RT_{approvalresponse}, RT_{emailservice}, \\
&RT_{employee}, RT_{emailrules}, RT_{approvalportal}, RT_{emailmessage}, RT_{emailresponse}\ \}.
\end{aligned} \tag{19}$$

Table 20 provides the description of each resource of the WCS. Table 21 provides the input and output resources for each task.

TABLE 20
RESOURCES FOR THE WCS TASK SYSTEM MODEL

Resource	Description
$RT_{weatherservice}$	Indicates the weather services used for the task.
$RT_{automatedservice}$	Indicates the automated service that invokes the weather service.
$RT_{weatherparameters}$	Indicates the input parameters of the weather service. The parameters could be stored in a database or configuration file and retrieved by the automated service when needed.
$RT_{weatherreport}$	The output of the weather report that is an XML document.
$RT_{employee}$	The employee performing the approval task.
$RT_{employeerole}$	Indicates the role of the employee performing the task.
$RT_{employeeknowledge}$	The employee's knowledge that plays a part in the approval task.
$RT_{approvalportal}$	The portal with which the user performs the approval process.
$RT_{emailrules}$	The email rules dictate the minimum condition on which the emails should be sent.
$RT_{approvalresponse}$	The employee's response indicating whether an email should be sent or not.
$RT_{emailmessage}$	The message of the emails sent to the user.
$RT_{emailservice}$	The email service that is used to send emails.
$RT_{emailresponse}$	The confirmation response of the email service.

TABLE 21
INPUT/OUTPUT SETS FOR THE WCS TASK SYSTEM MODEL

Tasks	Input set I_T	Output Set O_T
Weather	$RT_{weatherservice}$, $RT_{automatedservice}$, $RT_{weatherparameters}$	$RT_{weatherreport}$
Approval	$RT_{employee}$, $RT_{employeeknowledge}$, $RT_{approvalresponse}$, $RT_{emailrules}$, $RT_{weatherreport}$, $RT_{approvalportal}$, $RT_{emailmessage}$, $RT_{employeerole}$	$RT_{emailmessage}$, $RT_{weatherreport}$, $RT_{approvalresponse}$
Email	$RT_{emailmessage}$, $RT_{emailservice}$, $RT_{weatherreport}$	$RT_{emailresponse}$

Now, the composite $P^2_{FRAMEWORK}$ guidance is applied, classifying the WCS resources based on the six types of semantics: KS, ROS, RUS, UPS, INS, and COS.
Table 22 shows the classification of the resources based on the six types of semantics.

From Table 22, the four service-agents of the WCS are identi-
fied: $RT_{SA\_weatherservice}$, $RT_{SA\_automatedservice}$, $RT_{SA\_emailservice}$, and $RT_{SA\_employee}$. The service-
agents $RT_{SA\_weatherservice}$, $RT_{SA\_automatedservice}$, and $RT_{SA\_emailservice}$ represent mechanistic resources. The WCS example also includes a human interaction activity, which is the
task of approving or rejecting the sending of emails. The task is carried out by an employee $RT_{employee}$ of a certain role $RT_{employeerole}$ in the enterprise. The service-agent
$RT_{SA\_employee}$ represents the human activity.

TABLE 22
SEMANTIC-DIMENSION-BASED CLASSIFICATION OF THE WCS

Semantics	Resources		
KS	$RT_{employeeknowledge}$		
RUS	$RT_{emailrules}$		
ROS	$RT_{employeerole}$		
UPS	$RT_{employee}$		
INS	$RT_{SA\_weatherservice}$, $\quad RT_{SA\_automatedservice}$, $\quad RT_{SA\_emailservice}$, $RT_{SA\_employee}$		
COS	$RT_{weatherparameters}$, $RT_{weatherreport}$, $RT_{emailmessage}$, $RT_{emailresponse}$		

The advantage of the composite $P^2_{FRAMEWORK}$ approach is that the service-agents can be used to incorporate the necessary semantics into the executable process systematically from the abstract process models. The use of service-agents also provides a flexible approach for changing the semantics of the process because the semantics are integrated in a non-invasive manner as properties of agents.

The input and outputs sets are reclassified based on the identified service-agents and the classification of the resources according to the semantics. The reclassified input and output sets of the WCS are shown in Table 23.

TABLE 23

RECLASSIFIED INPUT/OUTPUT SETS FOR THE WCS TASK SYSTEM MODEL

Tasks	Input set I_T			Output Set O_T
	Service-Agents	COS	Associated Semantics	COS
Weather	$RT_{SA\_automatedservice}$, $RT_{SA\_weatherservice}$	$RT_{weatherparameters}$	NA	$RT_{weatherreport}$
Approval	$RT_{SA\_employee}$	$RT_{weatherreport}$, $RT_{approvalresponse}$, $RT_{emailmessage}$	$RT_{employeeknowledge}$, $RT_{employee}$, $RT_{employeerole}$, $RT_{emailrules}$, $RT_{approvalportal}$, $RT_{weatherreport}$	$RT_{emailmessage}$, $RT_{weatherreport}$, $RT_{approvalresponse}$
Email	$RT_{SA\_emailservice}$	$RT_{emailmessage}$, $RT_{weatherreport}$	NA	$RT_{emailresponse}$

The concept maps for the WCS tasks are depicted in Fig. 31 of the weather task, Fig. 32 of the approval task, and Fig. 33 of the email task. The concept maps show the representation of each task with the CPP semantics and also articulate the relationship of the CPP semantics to the service-agents.

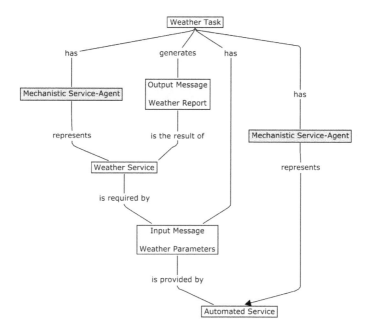

Fig. 31. Concept map of weather task in the WCS.

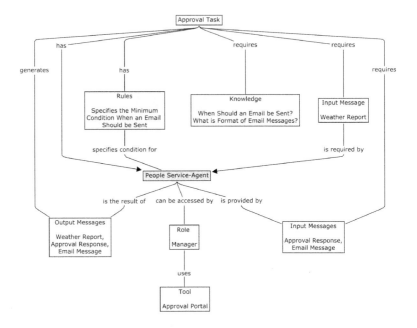

Fig. 32. Concept map of approval task in the WCS.

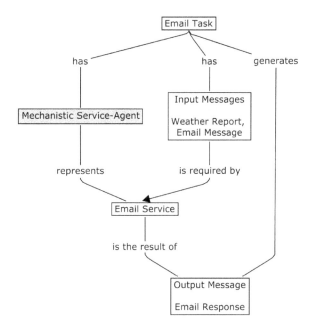

Fig. 33. Concept map of email task in the WCS.

An implementation of the WCS using BPEL is described to illustrate a potential mapping strategy between the abstract process model of the WCS (Table 23), and the BPEL executable processes. The mapping can be achieved either manually or automatically. The WCS is implemented using Intalio's business process management system (BPMS) system [66], which uses BPMN constructs [66] to derive the BPEL executable process. Therefore, the mapping to the BPEL processes is achieved in two steps: first, by mapping the service-agents to the BPMN constructs, and second, by deriving the BPEL processes from the BPMN model. Fig. 34 shows the WCS implemented using the Intalio BPMN designer. The human interactions tasks of the WCS is implemented as a people activity that is defined in BPEL4People specification [212]. Appendix D shows the deployed example of the WCS.

BPMN separates the different participants of a process using the pools indicated in Fig. 34 with large, rectangular boxes. Fig. 34 shows that there are five participants in the WCS: The automated service indicated by the "Interface" pool, the processes flow indicated by the "Process" pool, the "User" pool, the "WeatherSer-vice" pool, and the "EmailService" pool. Therefore, the different service-agents can be mapped to the BPMN pools. In the WCS example, $RT_{SA\_weatherservice}$ would be mapped to the "Weatherservice" pool, $RT_{SA\_automatedservice}$ would be mapped to the "WeatherService" pool, $RT_{SA\_emailservice}$ would be mapped to the "EmailService" pool, and $RT_{SA\_employee}$ would be mapped to the "User" pool. The tasks in BPMN are represented as small, rectangular boxes with curved edges. A gateway is represented as a diamond-shaped box with a cross. A gateway determines traditional decisions, as well as the forking, merging, and joining of paths. An if-loop in programming is a type of gateway. BPMN defines four attributes to describe user tasks:

- The performers attribute describes the human resources assigned to the task, which could be a user or a group of users.
- The inmessage attribute specifies the input messages for the task.
- The outmessage attribute specifies the output messages for the task.
- The implementation attribute describes the technologies that are used to perform the task.

In Fig. 34, the pool that encloses the human activity, the user pool in this example, is associated with the employee role.

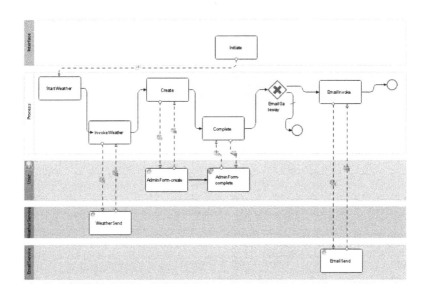

Fig. 34. BPMN model of the WCS.

Fig. 35 depicts a portion of the relevant syntax of the derived BPEL from the BPMN model. The human activity and the mechanistic services in the process are modeled as partner links in the BPEL model. We can also see that XForms are used to implement $RT_{approvalportal}$. Xforms provides an XML-based approach for developing the Web-based approval portal [66], [294]. The user role $RT_{employeerole}$ has been integrated using the NIST role-based access control [273]. The Intalio BPMS system provides support for handling human interaction. $RT_{emailrules}$ has been supported through the use of a gateway in the BPMN model, which is represented as <bpel:if> and <bpel:else> in the derived BPEL code. Appendix D shows a full listing of the derived BPEL code.

```
<bpel:partnerLinks>
    <bpel:partnerLink name="wFmagic_v0TFUGzDEdy8_fgUbrSKpgPlkVar"
partnerLinkType="diag:WFmagic_v0TFUGzDEdy8_fgUbrSKpg" initializePartnerRole="false"
myRole="Process_for_UserForThePortTypexformProcess" partnerRole="User_for_ProcessForXformPort" />
        ...
</bpel:partnerLinks>

  <xform:userOwner />
  <xform:roleOwner>examples\employee</xform:roleOwner>
  <xform:claimAction>
    <xform:user></xform:user>
    <xform:role></xform:role>
  </xform:claimAction>
  <xform:revokeAction>
    <xform:user>  </xform:user>
    <xform:role></xform:role>
  </xform:revokeAction>

        ...

  <xform:formUrl>oxf://PhDDemoCWS/AdminForm/AdminForm.xform</xform:formUrl>

   <xform:userProcessCompleteSOAPAction>http://example.com/AdminForm/AdminForm/xform/Process/
notifyTaskCompletion</xform:userProcessCompleteSOAPAction>
    <xform:isChainedBefore>
    </xform:isChainedBefore>

   <xform:userProcessEndpoint>http://localhost:8080/ode/processes/PhDDemoCWS/CompositeProcess/XWS/
Process/User/Process_for_UserForThePortTypexformProcessPort</xform:userProcessEndpoint>
    <xform:userProcessNamespaceURI>http://example.com/AdminForm/AdminForm/xform</
xform:userProcessNamespaceURI>
```

Fig. 35. BPEL code for the BPMN model of the WCS.

F. Validation and Verification of Process-Personalized Composite Services

The ultimate goal of validation is making sure that the "right" system is developed for the end-user with respect to meeting real-world needs, whereas the goal of verification is making sure that the system was developed "right" with respect to meeting the engineering requirement specifications [295]. V&V smoothes the transition between requirements and design by providing methods for evaluating the ability of a given approach to satisfy demanding technical requirements [296]. Together, V&V encompass the testing, analysis, demonstration, and examination methods used to determine whether a proposed design will satisfy system requirements.

This book supports V&V inherently. The motivation of this book is to reduce the process gap in composite services development. Process gap, as defined, is the gap between the user needs and implemented services. The process gap is reduced by incorporating the semantics of user needs for CPP, ensuring that the right, validated system is developed.

The integration of semantics also provides support for enhanced process analysis and composition of services, ensuring that a verified system is developed. Each of the three layers of the framework has feedback built in. For the composite services layer, feedback is achieved by two means. First, an appropriate process modeling approach, such as the task system model, provides an approach for analysis and verification of the models [297]. Second, by process enactment and mining, the run time behavior of the system can be verified. An interaction-pattern-based process mining approach is described in Chapter III and demonstrated in Section H of this chapter. The V&V support of the composite $P^2_{FRAMEWORK}$ is demonstrated using the two case study examples. Both the case studies demonstrate the integration of the different types of CPP semantics in composite services development for reducing process gap.

G. A Composite Process-Personalization Development Model

This section describes a development model (P^2_{DM}) for composite services with CPP. Fig. 36 depicts the P^2_{DM}. The P^2_{DM} is developed based on current SOA development models. The difference is that the CPP Management System (P^2_{EMS}) is added in the process management layer. Fig. 36 depicts the data layer at the bottom of the P^2_{DM}. The data layer can comprise such entities as databases, external service, XML documents, servers, tools, and portlets. Fig. 36 depicts the service layer above the data layer. The service layer consists of services that provide an interface to access the data layer. At the top of Fig. 36 is the interaction layer. Users can interact with the system through different means, including using tools, Web sites, desktop-based system, or pervasive computers. Fig. 36 depicts the process management layer below the interaction layer. The process management layer consists of executable processes and the P^2_{EMS}. The P^2_{EMS} is configured with CPP semantics. In the middle, below the process management layer and above the service layer, Fig. 36 depicts a composition engine that is responsible for the composition of services based on process descriptions. The link between the executable processes and the P^2_{EMS} is the service-agent.

During the development of a process model, an enterprise developer will link a service-agent to a resource. He will then assign properties to the service-agent. For example, the developer will assign the service-agent to be accesible only for certain

106

roles of users. The developer will also assign such user preferences to the service-agent that can indicate the type of tools, interfaces, and information that a user needs to complete the task. He will also assign relevant business rules and enterprise knowledge of a task to the service-agent. The composition engine will then handle the development of composite services. The P^2_{EMS} provides the environment for the support of the human interation with processes.

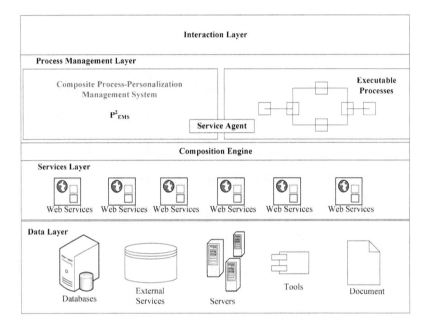

Fig. 36. A composite process-personalization development model.

H. Interaction Pattern Mining Example

The interaction-pattern-based process mining approach is demonstrated using the WCS example. The set of tasks of the WCS is

$$\tau_{WCS} = \{\text{Weather, Approval, Email, Reject}\}. \tag{20}$$

107

Table 23 provides the resources of the WCS for each task and shows the messages that are passed between the subsystems. The service-agents are used for representing the interacting subsystems of the WCS. The interaction-pattern-based process mining approach is concerned with message transfer between the service-agents.

The four service-agents are represented as $X1$, $X2$, $X3$, and $X4$. The states of the service-agents can be of three types: receiving, replying, and inactive, which are empirically represented as 1, 2, and 3, respectively. An input set of 1000 primary variables is generated with some standard time increment. The partial input set of the first twenty variables is provided in Table 24. The calculated normalized transmission t_{ij} is shown in Table 25. Appendix B shows the Web-based system for calculating the normalized transmission t_{ij}. Fig. 37 depicts the bar chart model of the interaction patterns between the different tasks of the WCS.

TABLE 24
PARTIAL INPUT SET FOR THE WCS EXAMPLE

j\i	1	2	3	4	5	6	7	8	9	10	11	12	13	14	15	16	17	18	19	20
X1	1	3	2	1	1	1	1	2	3	3	3	3	3	3	3	1	2	2	2	3
X2	3	2	1	2	2	1	3	1	3	3	3	3	1	3	1	2	1	3	2
X3	3	3	1	2	2	3	1	1	2	2	2	3	2	2	1	1	2	1	2	2
X4	2	2	1	1	3	1	1	2	1	2	1	1	3	3	1	2	3	2	3	2

TABLE 25
CALCULATED t_{ij} FOR THE WCS EXAMPLE

'i\j'	X1'	X2'	X3'	X4'
X1	0.0013	0.0014	0.0010	0.0020
X2	0.0015	0.0019	0.0010	0.0045
X3	0.0003	0.0014	0.0012	0.0008
X4	0.0014	0.0020	0.0019	0.0006

108

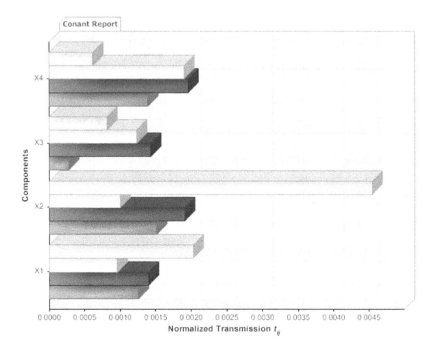

Fig. 37. Bar chart model of the interaction patterns.

The interaction patterns can also be modeled by drawing arrows that represent the interactions between the service-agents as shown in [62]. For convenience, any t_{ij} value less than or equal to 0.0010 is ignored. The internal communications of the service-agents are also ignored.

Table 25 is now modified as shown in Table 26. The interaction pattern model depicted in Fig. 38 shows the message transfer or interactions between the service-agents. One approach for identifying areas of optimization is by fixing a threshold value for t_{ij}. Any value above the threshold, for instance, might indicate a potential for bottleneck. For example, if the threshold value in this example is 0.0030, the interaction between X2 and X4 might be an area of concern. Further analysis would be necessary to assess the level of concern. The analysis would be aided by the semantics incorporated in the service-agents. Examples of analysis of composite services systems using Conant's decomposition approach is provided in [14], [15], [19], [287], [298].

TABLE 26
MODIFIED t_{ij} TABLE FOR THE WCS EXAMPLE

'I\j'	X1'	X2'	X3'	X4'
X1		0.0014		0.0020
X2	0.0015			0.0045
X3		0.0014		
X4	0.0014	0.0020	0.0019	

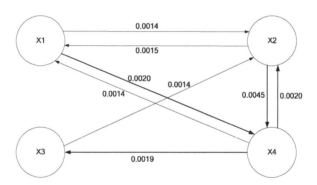

Fig. 38. Interaction pattern model depicting message transfer.

I. Case Study – Composite Process for Epidemiology Research

This section describes the GLIP case study, which demonstrates and validates the guidance offered by the composite $P^2_{FRAMEWORK}$. The case study is the GLIP stage of the composite process for epidemiology research depicted in Fig. 2 in Chapter I. The case study was developed in collaboration with Dr. Arnett and Dr. Vaughan of the Department of Epidemiology at UAB. A part of the overall the process is depicted using the role activity diagram in Fig. 39, which groups the set of tasks from the per-

110

spective of the roles of users interacting with the system. The RAD modeler software from instream services was used to depict the role activity diagram [66]. The role activity diagram depicts four types of users participating in the process: The researcher who directs the study, the research assistant who aids the researcher in data collection, the data analyst who performs the data analysis, and the post-doc who performs the GLIP. The researcher starts the study by designing the experiment. The researcher interacts with the data analyst to collect and store the results of the experiments. A database such as Microsoft Access is used to store the data. The assistant forwards the data to the analyst, who performs the analysis. The analyst uses such tools as Microsoft Excel, or SAS software to perform the analysis.

The GLIP is also depicted using the UML sequence diagram in Fig. 40. The sequence diagram is used to show the different tools, services, and other resources used in the GLIP. The sequence diagram also depicts the different knowledge input, rules, and messages that are part of the process.

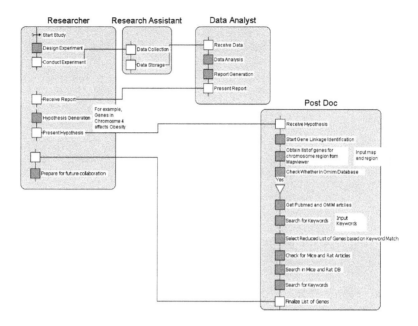

Fig. 39. Role activity diagram for composite epidemiology process.

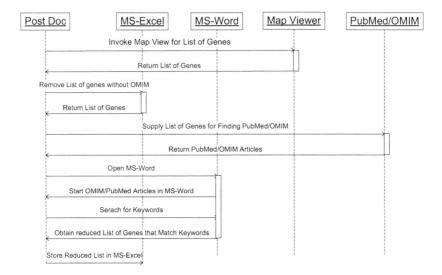

Fig. 40. Sequence diagram for the GLIP.

The results of the data analysis are presented to the researcher, who generates the hypothesis. The hypothesis in this case is that genes in human chromosome 4 play a critical role in causing obesity [49]. The researcher then presents the hypothesis to the post-doctoral fellow. The post-doctoral fellow then starts the GLIP. Table 27 outlines the steps in the GLIP [20].

TABLE 27
STEPS IN THE GLIP

Steps	Discussion
Step 1	The researcher visits the NCBI Web site and uses a tool called Map Viewer. Map Viewer shows integrated views of chromosome maps for many organisms, including human and numerous other vertebrates, invertebrates, fungi, protozoa, and plants [50]. The researcher uses the Map Viewer to obtain the chromosome maps and then to study the genes in a specific region of the chromosome. After making a series of selections through several Web pages, the researcher can select the specific region of the chromosome under investigation and also a few other details on the viewing style. The tool finally produces a page from which the genes on the sequence in the selected region of the chromosome can be downloaded. This downloaded document is a tab-delimited file and is most readable in spreadsheet applications, such as Microsoft Excel.
Step 2	On studying this document downloaded from Map Viewer, the researcher identifies the genes that have associated OMIM information. OMIM is a NCBI database that lists all the known diseases and the related genes in the human genome [50]. The researcher then manually goes to the NCBI OMIM search interface and looks up the OMIM information for each gene.
Step 3	Similarly, each gene in the list can be associated with one or multiple PubMed documents. PubMed is an NCBI database that stores citations for biomedical articles [50]. PubMed can be accessed via PubMed central or through Entrez. Entrez is a more sophisticated Web portal/search engine than NCBI tools that can be used to query many of the databases offered by NCBI. When the researcher types a gene into the search interface of PubMed, it displays a list of all documents from the database that cites that particular gene. Each document will have summaries or abstracts of the article.
Step 4	The OMIM and the PubMed information associated with each gene, offers vital information for the researcher to identify candidate genes for further research. This is done by carrying out a key-word search on the retrieved PubMed and OMIM information The genes that yield "hits" during the search are those genes that are identified as candidate genes. Other genes are discarded.
Step 5	The next step in the research is to find the evidence of the selected genes or genes with similar functions in rats or mice. Genes in rats or mice that show similar functions associated with the enlargement of the left ventricle, for example, would be evidence of linkage. The identified genes of rat and mouse are studied further. For this purpose several other databases from NCBI [50] such as Single Nucleotide Polymorphism, Taxonomy, and Nucleotide would be required. Other databases such as the Rat Genome Database [299] or the Mouse Genome Database [300] are also used. These databases have their individual interfaces or can be accessed through Entrez.

Note: Adapted from "Design of a Service-Oriented Composite Dashboard," by G. Sundar, R.S. Sadasivam, and M.M. Tanik, 2007, Proc. Int. Design and Process Technology, Antalya, Turkey, pp. 226-233. Copyright 2007 by the Society of Design and Process Science. Adapted with permission.

The goal of the GLIP is to identify the list of genes that can be studied further in other organisms, such as mice or rats. The description of the GLIP is listed in Table 27. The GLIP is modeled using the task system model (Fig. 41).

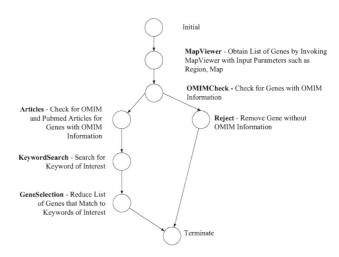

Fig. 41. Task system model of the GLIP.

The task system of the GLIP is

$$\tau_{GLIP} = \{(\ MapViewer, OMIMCheck \ ),(\ OMIMCheck, Articles \ ),$$
$$(\ Articles, KeywordSearch \ ),(\ KeywordSearch, GeneSelection \ )\}. \quad (21)$$

The set of resources of the GLIP is defined as

$$\rho_{GLIP} \equiv \{ \ RT_{postdocrole}, RT_{postdocuser}, RT_{mapviewer}, RT_{mapviewerinput},$$
$$RT_{postdocknowledge}, RT_{genes}, RT_{msexcel}, RT_{omimrule}, \quad (22)$$
$$RT_{pubmeddb}, RT_{omimdb}, RT_{articles}, RT_{msword}, RT_{searchkeyword} \ \}.$$

Table 28 shows the description of the resources of $C_{GLIP} = (\tau, < *)$. Table 29 shows the description of the resources of $C_{GLIP} = (\tau, < *)$. The resources are classified as shown in Table 30 according to the CPP semantics. The INS of the GLIP tasks is further classified as mechanistic services that can be composed and tools that the user needs for completing the process. The composable services are $RT_{pubmeddb}$ and

RT_{omimdb}. The tools that the user needs are $RT_{mapviewer}$, $RT_{msexcel}$, and RT_{msword}. The different service-agents of the GLIP are as follows: RT_{SA_omimdb}, $RT_{SA\_pubmeddb}$, and $RT_{SA\_postdoc}$. The reclassified input and output sets of the C_{GLIP} are shown in Table 31. Fig. 42 depicts the concept map of the Map Viewer task, Fig. 43 depicts the concept map of the OMIM check task, Fig. 44 depicts the concept map of the articles task, Fig. 45 depicts the concept map of the keyword search task, and Fig. 46 depicts the concept map of the gene selection task. The concept maps depict the relationship between the CPP and the service-agents for each task. Appendix E provides the six types of semantics captured for each task.

TABLE 28
RESOURCE INFORMATION OF THE GLIP

Resource	Description
$RT_{postdocrole}$	Represents the role of the user that participates in the process.
$RT_{postdocuser}$	Represents the user that is associated with the role. It also indicates the user's preference of tools for interacting with the process.
$RT_{mapviewer}$	Represents a tool that the user uses for obtaining the list of genes.
$RT_{mapviewerinput}$	Represents the input to the Map Viewer.
$RT_{postdocknowledge}$	Represents the knowledge of the user for providing appropriate input to the Map Viewer and appropriate keywords for identifying genes of interest for further studies.
RT_{genes}	Represents the list of genes that is obtained from the Map Viewer tool.
$RT_{msexcel}$	Represents the tool that the user uses for storing genes.
$RT_{omimrule}$	Represents an initial condition for selecting genes for which PubMed and OMIM articles must be obtained.
$RT_{pubmeddb}$	Represents the database that the user uses for obtaining PubMed articles.
RT_{omimdb}	Represents the database that the user uses for obtaining OMIM articles.
$RT_{articles}$	Represents the articles that are obtained from the PubMed and OMIM databases.
$RT_{fileupload}$	Represents the tool that the user uses to upload the list of genes for the articles task.
$RT_{articlesrule}$	Represents the NCBI rules for accessing the PubMed and OMIM databases.
RT_{msword}	Represents the tool that the user uses for organizing the articles.
$RT_{searchkeyword}$	Represents the keyword of interest that the user uses for selecting the list of genes for further studies.
$RT_{keywordresults}$	Represents the results of the keyword of interest search.
$RT_{geneselectionrule}$	Represents the rules for selecting the list of genes based on the results of the keyword of interest search.

TABLE 29
INPUT/OUTPUT RESOURCES OF THE GLIP

Tasks	Input Set I_T	Output Set O_T
MapViewer	$RT_{mapviewerinput}$, $RT_{mapviewer}$, $RT_{postdocuser}$, $RT_{postdocrole}$, $RT_{postdocknowledge}$	RT_{genes}
OMIMCheck	$RT_{omimrule}$, RT_{genes}, $RT_{msexcel}$, $RT_{postdocuser}$, $RT_{postdocrole}$, $RT_{postdocknowledge}$	RT_{genes}
Articles	RT_{omimdb}, $RT_{pubmeddb}$, RT_{genes}, $RT_{postdocuser}$, $RT_{postdocrole}$, $RT_{articlesrule}$, $RT_{fileupload}$	$RT_{articles}$
KeywordSearch	$RT_{articles}$, RT_{genes}, RT_{msword}, $RT_{searchkeyword}$, $RT_{postdocuser}$, $RT_{postdocrole}$, $RT_{postdocknowledge}$	$RT_{keywordresults}$
GeneSelection	RT_{genes}, $RT_{msexcel}$, $RT_{keywordresults}$, $RT_{geneselectionrule}$, $RT_{postdoc}$, $RT_{postdocuser}$, $RT_{postdocrole}$, $RT_{postdocknowledge}$, RT_{msword}	RT_{genes}

TABLE 30
SEMANTIC-DIMENSION-BASED CLASSIFICATION OF THE GLIP

Semantics	Resources
KS	$RT_{postdocknowledge}$
RUS	$RT_{omimrule}$, $RT_{articlesrule}$, $RT_{geneselectionrule}$
ROS	$RT_{postdocrole}$
UPS	$RT_{postdocuser}$
INS	$RT_{mapviewer}$, RT_{omimdb}, RT_{msword}, $RT_{pubmeddb}$, $RT_{msexcel}$, $RT_{fileupload}$
COS	$RT_{mapviewerinput}$, RT_{genes}, $RT_{searchkeyword}$, $RT_{keywordresults}$

TABLE 31
RECLASSIFIED INPUT/OUTPUT RESOURCES OF THE GLIP

Tasks	Input Set I_T			Output Set O_T
	Service-Agent	COS	Associated Semantics	COS
Map Viewer	$RT_{SA\_postdoc}$	$RT_{mapviewerinput}$	$RT_{mapviewer}$, $RT_{postdocrole}$, $RT_{postdocuser}$ $RT_{postdocknowledge}$	RT_{genes}
OMIM Check	$RT_{SA\_postdoc}$	RT_{genes}	$RT_{omimrule}$, $RT_{postdocrole}$, $RT_{postdocuser}$, $RT_{msexcel}$ $RT_{postdocknowledge}$	RT_{genes}
Articles	$RT_{SA\_postdoc}$ RT_{SA_omimdb} $RT_{SA\_pubmeddb}$	RT_{genes}	$RT_{postdocrole}$, $RT_{postdocuser}$, RT_{msword}, $RT_{searchkeyword}$, $RT_{pubmeddb}$, RT_{omimdb}, $RT_{postdocuser}$, $RT_{postdocrole}$	$RT_{articles}$
Keyword Search	$RT_{SA\_postdoc}$	RT_{genes} $RT_{searchkeyword}$ $RT_{articles}$	RT_{msword}, $RT_{postdocknowledge}$, $RT_{postdocrole}$, $RT_{postdocuser}$	$RT_{keywordresults}$
Gene Selection	$RT_{SA\_postdoc}$	RT_{genes}, $RT_{keywordresults}$	$RT_{msexcel}$, $RT_{postdocrole}$ $RT_{postdocknowledge}$, $RT_{postdocuser}$, RT_{msword}, $RT_{keywordresults}$, $RT_{geneselectionrule}$,	RT_{genes}

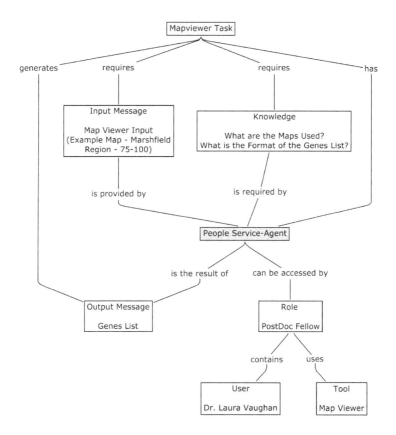

Fig. 42. Concept map of MapViewer task in the GLIP.

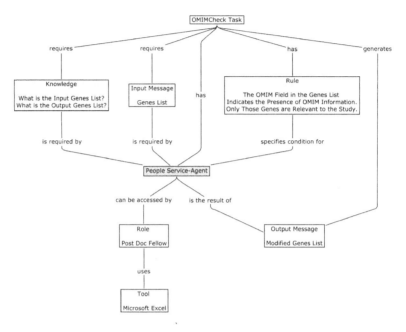

Fig. 43. Concept map of OMIMCheck task in the GLIP.

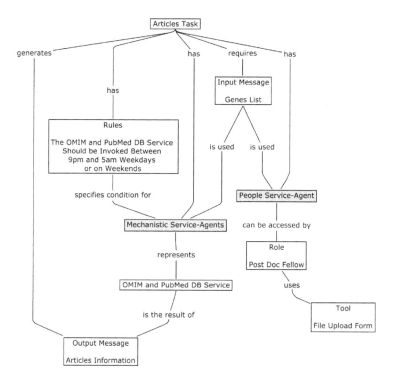

Fig. 44. Concept map of Articles task in the GLIP.

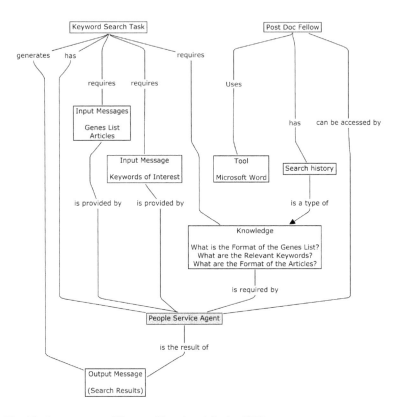

Fig. 45. Concept map of KeywordSearch task in the GLIP.

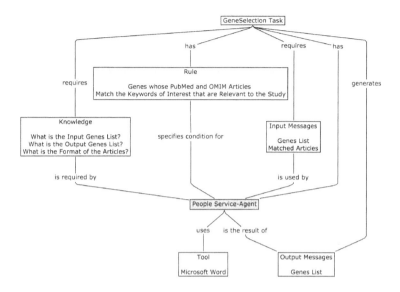

Fig. 46. Concept map of GeneSelection task in the GLIP.

J. Conclusion

In this chapter, the dimensions required to address the CPP challenge were defined. The composite $P^2_{FRAMEWORK}$, an architecture framework for guiding composite service development with emphasis on CPP, was described. The service-agent model was also described. Two case studies that demonstrate the composite $P^2_{FRAMEWORK}$ guidance were provided. A CPP development model was also described. The process enactment and mining approach was demonstrated.

V.   CONCLUSION AND FUTURE WORK

A. Summary

In this section, concluding remarks of this book are provided.

1) Motivation and Approach

Large systems development must address two types of integration: integration of mechanistic processes and integration of human interaction with processes [2]. Currently, integration efforts are mostly geared towards integration of mechanistic processes [34]. The CPP challenge is the lack of composite, integrated, and personalized support for human interaction with processes. As a result, the process gap occurs between the user needs and the implemented composite services. The large systems integration issues are discussed in the context of an enterprise. Two problems are identified for the lack of CPP in composite services:

- The first problem is the inability to capture the semantics of user needs comprehensively during the abstract process modeling of large systems. The loss of semantics in the abstract process modeling stage can be attributed to multiple reasons. One reason for the loss of semantics could be that the abstract process approach we choose might not have the capability to capture the semantics of user needs. This problem could be avoided by choosing an abstract process modeling approach that has better support for capturing the semantics of user needs, such as the task system model. Another reason for the loss of semantics could be that the developer's understanding of the problem domain, and the scope of integration of user needs, might not be sufficient to capture all the necessary semantics for CPP. This problem can be averted by using a systematic approach for capturing the necessary semantics of user needs.

- The second problem is the inability of executable process modeling technologies to configure and manage the semantics that were captured in the modeling stage [34]. The loss of semantics occurs because of a lack of support for human interaction in current technologies.

In addressing these issues, the composite $P^2_{FRAMEWORK}$ was developed for systematically guiding the development of composite services with CPP. The composite $P^2_{FRAMEWORK}$ realizes large systems integration with CPP based on the CPP dimensions, which identify the semantic and syntactic aspects of CPP, and the service-agent model, an abstraction that provides a modeling approach for the development of large systems with CPP. The service-agent abstraction model supports both the semantic and syntactic dimensions of CPP. A concept map was developed that visualizes and articulates the relationship of the CPP semantics with the service-agents for modeling and developing large systems with CPP.

Two case studies were used to demonstrate and validate the guidance offered by the composite $P^2_{FRAMEWORK}$. The first case study is the weather composite process. The second case study is the GLIP part of a composite process in epidemiology research.

2) Background

This book leverages research on process-oriented composite services development. In Chapter II and III, an overview of process-oriented composite services was provided. A composite service can be characterized as a system that is composed of a variety of services to realize the needs of a large system. Processes provide the syntactic and semantic elements of composite services to represent the needs of a large system. A composite service life cycle consists of four stages: need analysis and workflow generation, service discovery and selection, service integration and composition, composition analysis, monitoring, and optimization. The different architectures for composite services development were discussed. An example of composite services development was provided using SOA and Web services. An overview of the agent concept was provided.

In Chapter III, the notion that a process comprises semantic and syntactic elements of composite services was discussed. The semantic elements capture the essence of an enterprise, such as representing the work to be accomplished, the role of the user or the user that performs the work, and the services required to accomplish the work. The syntactic elements provide the structure for the process. An approach to representing processes uses task and resources [28]. A task constitutes the unit of compositional activity in a process. The task is specified in terms of its external behavior, such as the input it requires, the output it generates, its action or function, and its execution time. The tasks of a process could use multiple resources, such as a service, software, or a human activity.

The process life cycle was described as having the three stages: process modeling, process composition, and process analysis. The three stages were identified based on the EM model for composite services development. The engineering activity of developing composite services consists of abstract and executable process models. Abstract process models provide support for the semantic and syntactic aspects of an enterprise [8]. The abstract process models provide a good model for process analysis and enactment. The executable processes provide an executable representation of the process for composing services. The executable processes must be directly composable on a composition engine and can represent processes that are either mechanistic, that is devoid of any human interaction, or supportive of human activities.

An enterprise was considered as having four components [25]: peopleware, composed of people; netware, composed of networks or the communication medium of people; software, which the people use for doing their work; and hardware, which the software sits on. A process engineering approach should capture all of the components of an enterprise to accurately capture the semantic and syntactic aspects of an enterprise. The different viewpoints of process modeling were compared. The different technologies for process modeling were also discussed. The different process-formalisms for process engineering of large systems, such as Petri nets, pi-calculus, cubic and control flow graphs, and the task system model were also discussed. The task system model was described further, as it was used to represent the abstract process model for the WCS and the GLIP case studies in Chapter IV. A classification of the different process-oriented composite service approaches was provided, as agent-based and non-agent-based. Non-agent-based efforts use a variety of means to describe processes and can use either process technologies such as UML or BPEL, or process formalisms such as Petri nets, or patterns of processes for composition. Hybrid efforts using a combination of these also exist. The focus of agent-based efforts is on automation. The agent-based efforts can be sub-classified as approaches that focus on automating the generation of the processes and approaches that focus on automating the service composition after the design of the processes. The processes can be described using one of the non-agent approaches.

3) Contribution

In this book, a systematic framework was developed for guiding the development of composite services with CPP. The framework's guidance is based on the three dimensions of CPP. It uses the service-agent model for configuring and managing the CPP semantics in composite services.

The three dimensions of CPP were described. The primary dimension is the semantic-dimension. The semantic-dimension describes CPP semantics and can be classified as six types: KS, RUS, ROS, UPS, INS, and COS. The two types of syntactic-dimension are the change-dimension and the automation-dimension. The change-dimension and the automation-dimension are supported by the semantic-dimension of CPP.

The service-agent model is an approach for incorporating CPP in current executable modeling approaches. It combines two concepts, services, and agents. The services concept provides an abstraction for modeling different types of resources, such as services, methods, and objects. The agents' concept provides support for integrating the semantics to support the characteristics of human interactions.

The service-agent model provides a systematic approach for integrating the semantics of user needs. In this model, human activity and mechanistic resources are represented as service-agents. The service-agent supports two types of composite services activity. First, using agents allows an analyst to enact the process and simulate the human interactions. The simulations can then be used to analyze and optimize the process.

Second, it provides support for the enhanced composition of services. The service-agent achieves this by providing a description of the composite process and its resources enhanced with the semantics of user needs. Therefore, the composition engine can use the semantics for a more precise discovery and selection of services that match the user needs.

The composite $P^2_{\text{FRAMEWORK}}$ has three layers: core competency layer, business model layer, and composite services layer. Each of these layers is associated with the CPP semantics. The core competency and business model layers provide a direction and scope for identifying the integration needs of the large systems. The focus of this book was on the composite services layer.

The WCS example was used to demonstrate the interaction pattern mining approach. The interaction pattern mining approach was described as an approach for process analysis and optimization.

A CPP development model based on the SOA model was also described. Two case studies were used to demonstrate and validate the composite $P^2_{\text{FRAMEWORK}}$ guidance. The second case study is the GLIP part of the composite process in epidemiology research. Both case studies demonstrated the guidance provided by the framework for developing systems with CPP.

B. Impact of Framework Guidance on Composite Process-Personalization

The characteristics of human interactions with processes were discussed in Chapter III. In this section, the impact of the systematic guidance of the composite $P^2_{\text{FRAMEWORK}}$ on addressing the needs of human interactions with processes is discussed. In Table 32, the impact of the framework guidance on the five challenges raised by Harrison-Broninski is described [34].

The four types of team distributions described by [236] can be addressed using RUS and UPS. RUS addresses the temporal and social distribution. It associates a task with a role rather than a user; therefore, if a user changes, the process does not have to change. A new user will then be associated to the role. In addition, according to their expertise, users can belong to different roles in an enterprise. By using the semantics provided by UPS, INS, and COS, composite services with CPP can personalize users' interactions with their tasks. Thus, spatial distribution and technological distribution of teams can be handled. Table 33 describes the impact of the impact of the framework guidance on the service functions characteristics described by Ramamoorthy [2].

TABLE 32

DIMENSION ADDRESSING CPP NEEDS POINTED BY HARRISON-
BRONINSKI

Needs	Discussion
Connection visibility	ROS provides a representation of the process participants and their interactions. UPS provides a representation of the users in the process. INS provides a representation of the resources that are involved on the process. COS provides a representation of the messages that are transferred during the process.
Structured messaging	KS provides the information that is necessary to interact with a process. RUS provides the rules that are necessary to make decision during the interactions with a process. INS provides a representation of the resources that are involved on the process. COS provides a representation of the messages that are transferred during the process. Thus, the composite services with CPP can support the interactions of the user in a more informed manner.
Support for mental work	As discussed above, the framework can provide an integrated and personalized information and knowledge environment using the semantic-dimension. An example of CPP support for mental work is the TAS model discussed in Chapter I.
Supportive rather than prescriptive activity management	The framework can address this challenge using multiple semantics: KS and RUS provide an information context to the composite service. ROS and UPS provide a representation of the user's role in the process and his preferences. INS and COS provide a representation of the resources and the messages that are needed to interact with the process. Thus, the composite service with CPP would be able to respond to the user's needs in a more informed manner.
Process change processes	The change-dimension addresses this need. Process will be continually monitored to optimize and change them if necessary. Leveraging composite services with CPP implies a flexible system development model that can accommodate changes. The enhanced semantics helps in analyzing and optimizing the process.

130

TABLE 33
DIMENSION ADDRESSING CPP NEEDS POINTED BY RAMAMOORTHY

Needs	Discussion
Human-needs-driven	The composite $P^2_{FRAMEWORK}$ can address this challenge using UPS. By describing the needs and preferences in the UPS, we can leverage customization and humanization technologies for supporting the user needs. In addition, by using KS and RUS, we can provide an integrated and personalized information and knowledge environment for the user.
Knowledge-intensive, high mental support	This challenge can be addressed in a similar manner to Harrison-Broninski's support for mental work challenge.
Automation-intensive to reduce manual effort	The composite $P^2_{FRAMEWORK}$ can address this challenge using the automation-dimension.
Human-interaction-intensive	The composite $P^2_{FRAMEWORK}$ can address this challenge using UPS, INS, and COS. By providing KS and RUS, the framework can enable the composite service to support the interactions of the user in a more informed manner.
Information-technology-intensive	The composite $P^2_{FRAMEWORK}$ can address this challenge by leveraging the semantics provided by UPS, INS, and COS. An example is the TAS model discussed in Chapter I. The change-dimension can also help by optimizing the process. In addition, automation can help to reduce the workload of the user.
Team-based	The composite $P^2_{FRAMEWORK}$ can address this primarily using ROS and UPS.

C. Future work

This book incorporates a broad area of research for developing a systematic approach for CPP. Several research opportunities exist in this area of CPP. First, six types of semantics of CPP were identified in this book. While several approaches can be used for representing the semantics in machine-accessible form, continued research is required to find effective ways of identifying, organizing, and representing the semantics of CPP.

Another area for research is process-oriented composite service development. The service-agent provides a way to enhance the description of an executable process with semantics. Continued research is required to enhance the composition of services leveraging the service-agent model. Other areas of composite services such as trust, security, integrity, and reliability must also be addressed to achieve the effective integration of systems on a global scale.

The third research area is human interaction with processes. As technology improves, our interaction with processes also changes. Continued research is required on human interaction with processes to take advantage of technological advancements in composite services development with CPP.

The fourth research area is process analysis and optimization. An interaction-pattern-based process mining approach is discussed in this book. Continued research is needed to take advantage of the process mining approach together with the CPP semantics for process analysis and optimization.

Another research area is the TAS model, which provides enhanced integration of processes. Integration of the TAS model with composite services requires that several issues be addressed, such as data transparency between the tools and availability of tools in a service form [17], [20], [21].

LIST OF REFERENCES

[1]   H.A. Simon, *The Sciences of the Artificial*. Cambridge, MA: The MIT Press, 2001.

[2]   C.V. Ramamoorthy, "A study of the service industry - functions, features, and control," *ICICE Trans. Commun.*, vol. E83-B, no. 5, pp. 885-903, May 2000.

[3]   C. Petrie and C. Bussler, "Service agents and virtual enterprises," *IEEE Internet Comput.*, vol. 7, no. 4, pp. 68-78, Jul.-Aug. 2003.

[4]   D.J. Mandell and S.A. McIlraith, "Adapting BPEL4WS for the semantic Web: The bottom-up approach to Web service interoperation," in *Proc. 2nd Int. Semantic Web Conf.*, Sanibel Island, FL, 2003, pp. 227-241.

[5]   P. Fingar, *Extreme Competition: Innovation and the Great 21st Century Business Reformation*. Tampa, FL: Meghan-Kiffer Press, 2006.

[6]   R. Gleghorn, "Enterprise application integration: a manager's perspective," *IT Professional*, vol. 7, no. 6, pp. 17-23, Nov.-Dec. 2005.

[7]   A.F. Vojdani, "Tools for real-time business integration and collaboration," *IEEE Trans. on Power Syst.*, vol. 18, no. 2, pp. 555-562, 2003.

[8]   S. Khoshafian, *Service-Oriented Enterprises*. Boca Raton, FL: Auerbach Publications, Taylor & Francis Group, 2007.

[9]   M.B. Juric, B. Mathew, and P. Sarang, *Business Process Execution Language for Web Services: BPEL and BPEL4WS*. Packt Publishing, 2004.

[10]   M. Harvey. (2005, Jul.). What is business process modeling [Online]. Available:
http://www.onjava.com/pub/a/onjava/2005/07/20/businessprocessmodeling.html

[11]   S. Stabb, "Web services: been there, done that," *IEEE Intell. Syst.*, vol. 18, no. 1, pp. 72-85, Jan.-Feb. 2003.

[12]   R.S. Sadasivam, U.J. Tanik, and M.M. Tanik, "A test-bed for the correlation center of digital services," in *Proc. IEEE SouthEastCon*, Columbia, SC, 2002, pp. 381 -386.

[13]   R.S. Sadasivam, "Design of a correlation center for digital services," M.S. thesis, Dept. Elec. and Comput. Eng., Univ. of Alabama at Birmingham, Birmingham, AL, 2002.

[14]   R.S. Sadasivam, "Dynamic reconfiguration of field-agents to process-agents in distributed systems," in *Proc. Int. Design and Process Technology*, Turkey, 2004, pp. 220-229.

[15]   A. Mamuwallah, R.S. Sadasivam, and M.M. Tanik, "A Simon-Conant based entropic approach to function point analysis," in *Proc. Int. Design and Process Technology*, Beijing, China, 2005, pp. 733-738.

[16]   R.S. Sadasivam, M.M. Tanik, L. Casebeer, D. Allison, J. Gemmill, J. Lynn, B. Bryant, Y.-F. Wu, M. Bieber, and L. Jololian, "Component-based approach for scientific services for education and research (scientific SEARCH)," in *Proc. 2nd Global Educ. Technology in Sci. Symp. (GETS)*, Univ. of Arkansas at Little Rock, Little Rock, AR, 2005, pp. 40-48.

[17]   R.S. Sadasivam, G. Sundar, M.M. Tanik, and M.N. Tanju, "Process personalization framework for service-driven enterprises," in *Proc. IEEE SouthEastCon*, Memphis, TN, 2006, pp. 159-164.

[18]   R.S. Sadasivam, M.M. Tanik, and R. Kristofco, "A service-oriented approach for intelligent information integration and retrieval," in *Proc. Int. Design and Process Technology*, San Diego, CA, 2006, pp. 558-565.

[19]   R.S. Sadasivam, R. Goli, R. Kathiru, A. Krishnan, Y. Tuncer, M.M. Tanik, and S. Thompson, "A service-based entropic model for sensors orchestration," *Int. J. of Distributed Sensor Networks,* to be published.

[20]   G. Sundar, R.S. Sadasivam, and M.M. Tanik, "Design of a service-oriented composite dashboard," in *Proc. Int. Design and Process Technology*, Antalya, Turkey, 2007, pp. 226-233.

[21]   R.S. Sadasivam, G. Sundar, M.M. Tanik, L. Jololian, and M.N. Tanju, "A process personalization model for enabling biological research," in *Proc. Int. Design and Process Technology*, Antalya, Turkey, 2007, pp. 168-174.

[22]   R.S. Sadasivam, L. Jololian, and M.M. Tanik, "Java 2 enterprise edition implementation of a distributed business," in *Proc. Int. Design and Process Technology*, Austin, TX, 2003, pp. 583-587.

[23]   R.S. Sadasivam, M.M. Tanik, J. Gemmill, and T. Jannett, "Cyberinfrastructure development – a component based approach with software agents," in *Proc. Int. Design and Process Technology*, Austin, TX, 2003, pp. 714-722.

[24]   R.S. Sadasivam, M.M. Tanik, and L. Jololian, "Drag and drop communication of data via a computer network," U.S. Patent, 2006.

[25]   M. Bieber, M. Bartolacci, J. Fjermestad, F. Kurfess, L. Qianhong, M. Nakayama, A. Rana, W. Rossak, R. Scherl, M. Tanik, J. Wang, R. Yeh, P. Ng, R. Sweeney, and F. Vitali, "Electronic enterprise engineering: an outline of an architecture," in *Proc. Int. Conf. and Workshop on Eng. of Comput.-Based Syst.*, Monterey, CA, 1997, pp. 376-383.

[26]   D.E. Atkins, K.K. Droegemeier, and S.I. Feldman, "Revolutionizing science and engineering through cyberinfrastructure," NSF, BRPCI Report, 2003.

[27]   F. Berman, G. Fox, and A.J.G. Hey, *Grid Computing: Making the Global Infrastructure a Reality*. New York, NY: Wiley, 2003.

[28]   E.G. Coffman and P.J. Denning, *Operating Systems Theory*. Englewood Cliffs, NJ: Prentice Hall, 1973.

[29]   S.F. Mills and M.M. Tanik, "Resource focused process engineering for the distributed enterprise," *Int. J. of Comput. Integrated Manufacturing*, vol. 13, no. 3, pp. 187-203, 2000.

[30]   S.F. Mills, "A resource-focused framework for process engineering," Ph.D. dissertation, Dept. Eng. and Appl. Sci., Southern Methodist Univ., Dallas, TX, 1997.

[31]   S. Delcambre, "A software process modeling framework as a basis for process analysis and improvement," Ph.D. dissertation, Dept. Eng. and Comput. Sci., Southern Methodist Univ., Dallas, TX, 1994.

[32]   S. Delcambre and M.M. Tanik, "Using task system templates to support process description and evolution," *J. of Syst. Integration*, vol. 8, no. 1, pp. 83-111, 1998.

[33]   D. Wu and M. Turoff. (2001). Personalization in user interface [Online]. Available: http://eies.njit.edu/~turoff/coursenotes/CIS732/samplepro/CIS732FinalProject_Dezhi.htm

[34]   K. Harrison-Broninski, *Human Interactions: The Heart and Soul of Business Process Management*. Tampa, FL: Meghan-Kiffer Press, 2005.

[35]   M. Castells, *The Internet Galaxy: Reflections on the Internet, Business, and Society*. UK: Oxford Univ. Press, 2003.

[36]   R.S. Sadasivam, U.J. Tanik, and M.M. Tanik, "A test-bed for the correlation center for digital services," Univ. of Alabama at Birmingham, Birmingham, AL, 2001-08-ECE-006, 2001.

[37]   P. Fingar, H. Kumar, and T. Sharma, *Enterprise E-Commerce: The Software Component Breakthrough for Business-to-Business Commerce*. Tampa, FL: Meghan-Kiffer, 2000.

[38]   A.M. Madni, "Thriving on change through process support: the evolution of ProcessEdge^{TM} enterprise suite and TeamEdge^{TM}," *Inform. Knowledge Syst. Manage.*, vol. 2, no. 1, pp. 7-32, 2000.

[39]   J.W.J. Gijsen, N.B. Szirbik, and G. Wagner, "Agent technologies for virtual enterprises in the one-of-a-kind-production industry," *Int. J. of Electron. Commerce*, vol. 7, no. 1, pp. 9, 2002.

[40]   H.P. Alesso and C.F. Smith, *Developing Semantic Web Services*. Natick, MA: A. K. Peters, 2005.

[41]   E. Newcomer and G. Lomow, *Understanding SOA with Web Services*. Upper Saddle River, NJ: Addison-Wesley, 2005.

[42]   J. Pasley, "How BPEL and SOA are changing Web services development," *IEEE Internet Comput.*, vol. 9, no. 3, pp. 60-67, 2005.

[43]   W.M.P. van der Aalst, L. Aldred, M. Dumas, and A.H.M. ter Hofstede, "Design and implementation of the YAWL system," in *Advanced Inform. Syst. Eng.*, 2004, pp. 142-159.

[44]   F. Leymann, D. Roller, and M.-T. Schmidt, "Web services and business process management," *IBM Syst. J.*, vol. 41, no. 2, 2002.

[45]   A. Stone, "Demanding Internet enterprise," *IEEE Internet Comput.*, vol. 8, no. 3, pp. 13-14, 2004.

[46]   W.T. Tsai, "Service-oriented system engineering: a new paradigm," in *Proc. IEEE Int. Workshop on Service-Oriented Syst. Eng.*, 2005, pp. 3-6.

[47]   R.T.-Y. Yeh, K. Pearlson, and G. Kozmetsky, *Zero Time: Providing Instant Customer Value – Every Time, All the Time*. New York, NY: Wiley, 2000.

[48]   P. Fingar. (2005, Aug.). Business process management: the next generation (a preliminary draft of a book chapter on extreme competition) [Online]. Available: http://www.bpcommunity.org/2005/08/01/bpm-the-next-generation/

[49]   J. Wu, M.A. Province, H. Coon, S.C. Hunt, J.H. Eckfeldt, D.K. Arnett, G. Heiss, C.E. Lewis, R.C. Ellison, D.C. Rao, T. Rice, and A.T. Kraja, "An investigation of the effects of lipid-lowering medications: genome-wide linkage analysis of lipids in the HyperGEN study," *BMC Genet.*, vol. 8, no. 1, pp. 60, Sept 10, 2007.

[50]   D.L. Wheeler, D.M. Church, R. Edgar, S. Federhen, W. Helmberg, T.L. Madden, J.U. Pontius, G.D. Schuler, L.M. Schriml, E. Sequeira, T.O. Suzek, T.A. Tatusova, and L. Wagner, "Database resources of the National Center for Biotechnology Information: update," *Nucleic Acids Res.*, vol. 32, database issue, pp. 35-40, Jan. 1, 2004.

[51]   D.J. Elzinga, T. Horak, L. Chung-Yee, and C. Bruner, "Business process management: survey and methodology," *IEEE Trans. on Eng. Manage.*, vol. 42, no. 2, pp. 119-128, 1995.

[52]   M. Havey, *Essential Business Process Modeling*. Sebastopol, CA: O'Reilly, 2005.

[53]   K.C. Desouza, Y. Awazu, and P. Baloh, "Managing knowledge in global software development efforts: issues and practices," *IEEE Softw.*, vol. 23, no. 5, pp. 30-37, 2006.

[54]   P. Fingar. (2005, Nov.). The coming IT flip flop: and the emergence of hu-
         man   interaction   management   systems   [Online].   Available:
         www.bptrends.com/publicationfiles/12-05ART-ComingITFlipFlop-
         Fingar.pdf

[55]   W. Tolone, B. Chu, J. Long, R. Willhelm, T. Finin, Y. Peng, and A.
         Boughannam, "Supporting human interactions within integrated manufactur-
         ing systems," *Int. J. of Advanced Manufacturing*, vol. 1, no. 2, pp. 221-234,
         Dec. 1998.

[56]   T. Andersson, I. Bider, and R. Svensson, "Aligning people to business proc-
         esses experience report," *Softw. Process Improvement and Practice*, vol. 10,
         pp. 403-413, 2005.

[57]   B. Chu, R. Wilhelm, W.J. Tolone, M. Hegedus, J. Fesko, T. Finin, Y. Peng,
         C. Jones, J. Long, M. Matthes, J. Mayfield, J. Shimp, and S. Su, "Integrating
         manufacturing softwares for intelligent planning-execution: A CIIMPLEX
         perspective," in *Proc. Int. Soc. for Optical Eng. SPIE*, 1997, pp. 96-116.

[58]   I.-S. Fan and C. Albert, "Process model for diverse stakeholders goals," in
         Proc. Workshop on Goal-Oriented Business Process Modeling (GBPM),
         London, 2002, pp. 1-6.

[59]   J.D. Novak and A.J. Cañas, "The theory underlying concept maps and how
         to construct them," Florida Inst. for Human and Machine Cognition, Tech.
         Rep.: IHMC CmapTools, 2006.

[60]   T. Keller and S.-O. Tergan, "Visualizing knowledge and information: an in-
         troduction," in *Knowledge and Information Visualization*. 2005, pp. 1-23.

[61]   A. Castro, P. Rocca-Serra, R. Stevens, C. Taylor, K. Nashar, M. Ragan, and
         S.-A. Sansone, "The use of concept maps during knowledge elicitation in on-
         tology development processes - the nutrigenomics use case," *BMC Bioinfor-
         matics*, vol. 7, no. 1, pp. 267, 2006.

[62]   R.C. Conant, "Detecting subsystems of a complex system," *IEEE Trans. on
         Syst., Sci. and Cybern.*, vol. 2, pp. 550-553, Sept. 1972.

[63]   R.C. Conant, "Information transfer in complex system: with applications to
         regulations," Ph.D. dissertation, Dept. Elec. Eng., Univ. of Illinois, IL, 1968.

[64]   P. Doshi, R. Goodwin, R. Akkiraju, and K. Verma, "Dynamic workflow
         composition using Markov decision processes," in *Proc. IEEE Int. Conf. on
         Web Services*, 2004, pp. 576-582.

[65]   The-Stencil-Group. (2002, Jul. 19). The evolution of UDDI [Online]. Avail-
         able: http://www.uddi.org/pubs/the_evolution_of_uddi_20020719.pdf

[66]   OMG. (2006, Feb.). Business process modeling specification: OMG final
         adopted specification [Online]. Available: http://www.bpmn.org/Documents/
         OMG%20Final%20Adopted%20BPMN%201-0%20Spec%2006-02-01.pdf

[67]   E.M. Maximilien and M.P. Singh, "Agent-based architecture for autonomic Web service selection," in *Proc. AAMAS*, Melbourne, Australia, 2003, pp. 293-307.

[68]   G. Glass. (2000). Applying Web services to applications [Online]. Available: http://www-106.ibm.com/developerworks/webservices/library/ws-peer1.html

[69]   J. Roy and A. Ramanujan, "Understanding Web services," *IT Professional*, vol. 3, no. 6, pp. 69-73, 2001.

[70]   A. Brown, S. Johnston, and K. Kelly. (2002, Nov.). A Rational software white paper: Using service-oriented architecture and component-based development to build service application. *developerWorks* [Online]. Available: http://www-106.ibm.com/developerworks/rational/library/content/03July/2000/2169/2169.pdf

[71]   K. Channabasavaiah, K. Holley, and E.M. Tuggle Jr. (2003, Dec.). Migrating to a service-oriented architecture, part 1. *developerWorks* [Online]. Available: ftp://www6.software.ibm.com/software/developer/library/ws-migratesoa.pdf

[72]   D. Booth, H. Haas, F. McCabe, E. Newcomer, M. Champion, C. Ferris, and D. Orchard. (2004, Feb.). Web services architecture. [Online]. Available: http://www.w3.org/TR/2004/NOTE-ws-arch-20040211/

[73]   A. Layman, "Web services framework," in *Proc. W3C Workshop on Web Services*, San Jose, CA, 2001.

[74]   World Wide Web Consortium. (2004). Web services consortium [Online]. Available: http://www.w3c.org

[75]   J.M. Myerson. (2002). Web services achitectures: how they stack up [Online].   Available: http://www.webservicesarchitect.com/content/articles/myerson01.asp

[76]   M. Paolucci. (2004, Feb.). DAML-S virtual machine [Online]. Available: http://www2003.org/cdrom/papers/alternate/P829/p829-paolucci_node10.xhtml

[77]   M. Luck, R. Ashri, and M. D'inverno, *Agent-Based Software Development*. Norwood, MA: Artech House, Inc., 2004.

[78]   C. Peltz, "Web services orchestration and choreography," *IEEE Comput.*, vol. 36, no. 10, pp. 46-52, 2003.

[79]   E. Miller, R. Swick, D. Brickley, J. Hendler, G. Schreiber, and D. Connoly. (2004). Semantic Web [Online]. Available: http://www.w3.org/2001/sw/

[80]   J. Heflin and J. Hendler, "A portrait of the semantic Web in action," *IEEE Intell. Syst.*, vol. 16, no. 2, pp. 54-59, Mar.-Apr. 2001.

[81]   S.A. McIlraith, T.C. Son, and H. Zeng, "Semantic Web services," *IEEE Intell. Syst.*, vol. 16, no. 2, pp. 46-53, Mar.-Apr. 2001.

[82]   T. Berners-Lee, J. Hendler, and O. Lassila. (2001, May). The semantic Web. *Scientific Amer.* [Online]. Available: http://www.sciam.com/article.cfm?articleID=00048144-10D2-1C70-84A9809EC588EF21

[83]   J. Euzenat, "Research challenges and perspectives of the semantic Web," *IEEE Intell. Syst.*, vol. 17, no. 5, pp. 86-88, Sept. 2002.

[84]   D. Martin, M. Burstein, G. Denker, Hobbs, L. Kagal, O. Lassila, D. McDermott, S. McIlraith, M. Paolucci, B. Parsia, T. Payne, M. Sabou, E. Sirin, M. Solanki, N. Srinivasan, and K. Sycara. (2004). OWL-S 1.0 release [Online]. Available: http://www.daml.org/services/owl-s/1.0/

[85]   F. Casati, S. Ilnicki, L. Jin, V. Krishnamoorthy, and M.-C. Shan, "Adaptive and dynamic service composition in eFlow," in *Proc. Conf. on Advanced Inform. Syst. Eng.*, 2000, pp. 13-31.

[86]   DanaNau, T.-C. Au, O. Ilghami, U. Kuter, J.W. Murdock, D. Wu, and F. Yaman, "SHOP2: an HTN planning system," *J. Artificial Intelligence Res.*, vol. 20, pp. 379-404, Dec. 2003.

[87]   F. Cao, "Model driven development and dynamic composition of Web services," Ph.D. dissertation, Dept. Comput. and Inform. Sci., Univ. of Alabama at Birmingham, Birmingham, AL, 2005.

[88]   N. Gupta, R.R. Raje, A.M. Olson, B.R. Bryant, M. Auguston, and C.C. Burt, "Analyzing the Web services and UniFrame paradigms," in *Proc. Southeastern Softw. Eng. Conf. (SESEC)*, Huntsville, AL, 2003.

[89]   T. Oinn, M. Addis, J. Ferris, D. Marvin, M. Senger, M. Greenwood, T. Carver, K. Glover, M.R. Pocock, A. Wipat, and P. Li, "Taverna: a tool for the composition and enactment of bioinformatics workflows," *Bioinformatics*, vol. 20, no. 17, pp. 3045-3054, Nov. 22, 2004.

[90]   T. Oinn, M. Greenwood, M. Addis, M.N. Alpdemir, J. Ferris, K. Glover, C. Goble, A. Goderis, D. Hull, D. Marvin, P. Li, P. Lord, M.R. Pocock, M. Senger, R. Stevens, A. Wipat, and C. Wroe, "Taverna: lessons in creating a workflow environment for the life sciences," in *Proc. Concurrency and Computation: Practice and Experience Grid Workflow*, 2005, pp. 1067-1100.

[91]   K. Wolstencroft, T. Oinn, C. Goble, J. Ferris, C. Wroe, P. Lord, K. Glover, and R. Stevens, "Panoply of utilities in taverna," in *Proc. 1ˢᵗ Int. Conf. on e-Sci. and Grid Computing*, 2005, pp. 156-162.

[92]   M. David and B. Pagurek, "A runtime composite service creation and deployment infrastructure and its applications in Internet security, e-commerce, and software provisioning," in *Proc. Comput. Softw. and Applicat. Conf.*, 2001, pp. 371-376.

[93]   D. Mennie and B. Pagurek, "An architecture to support dynamic composition of service components," in *Proc. 5th Int. Workshop on Component-Oriented Programming (WCOP)*, Sophia Antipolis, France, 2000.

[94]   R. de Knikker, Y. Guo, J.L. Li, A.K. Kwan, K.Y. Yip, D.W. Cheung, and K.H. Cheung, "A Web services choreography scenario for interoperating bioinformatics applications," *BMC Bioinformatics*, vol. 5, pp. 25, Mar. 10, 2004.

[95]   R.T. Fielding, "Architectural styles and the design of network-based software architectures," Ph.D. dissertation, Dept. Inform. and Comput. Sci., Univ. of California, Irvine, CA, 2000.

[96]   R. McMillan. (2003). A RESTful approach to Web services [Online]. Available: http://www.networkworld.com/ee/2003/eerest.html

[97]   S. Vinoski, "Putting the "Web" into Web services. Web services interaction models," *IEEE Internet Comput.*, vol. 6, no. 4, pp. 90-92, 2002.

[98]   J. McCarthy. (2002, Jun.). Reap the benefits of document style Web services [Online]. Available: http://www-128.ibm.com/developerworks/webservices/library/ws-docstyle.html

[99]   World Wide Web Consortium. (2006, Feb.). Naming and addressing [Online]. Available: http://www.w3.org/Addressing/

[100]   A. Akram, D. Meredith, and R. Allan, "Best practices in Web service style, data binding and validation for use in data-centric scientific application " in *Proc. eSci. All Hands Meeting*, UK, 2006, pp. 10.

[101]   A.P. Saygin, I. Cicekli, V. Akman, and J.H. Moor. (2000). Turing test: 50 years later [Online]. Available: http://cogprints.org/1925/

[102]   K.A. Gluck and R.W. Pew, *Modeling Human Behavior with Integrated Cognitive Architectures: Comparison, Evaluation, and Validation*. Mahwah, NJ: Lawrence Erlbaum, 2005.

[103]   S. Maarten and J.C. William, "Modeling and simulating work practice: a method for work systems design," *IEEE Intell. Syst.*, vol. 17, no. 5, pp. 32-41, Sept. 2002.

[104]   H.S. Nwana. (1996). Software agents: an overview [Online]. Available: http://www.sce.carleton.ca/netmanage/docs/AgentsOverview/ao.html

[105]   A. Kay, "Computer software," *Scientific Amer.*, vol. 251, no. 3, pp. 53-59, 1984.

[106]   M. Minsky and S. Papert, *Artificial Intelligence*. Oregon State System of Higher Education, 1974.

[107]   M. Minsky, *The Society of Mind*. MA: Simon & Schuster, Inc., 1988.

[108]   L. Gasser and M. Huhns, *Distributed Artificial Intelligence 2*. San Mateo, CA: Morgan Kaufmann, 1989.

[109]   B. Chaib-draa, B. Moulin, R. Mandiau, and P. Millot, "Trends in distributed artificial intelligence," *Artificial Intell. Review*, vol. 6, pp. 35-66, 1992.

[110]   L. Gasser, J.S. Rosenschein, and E. Ephrati, "Tutorial: introduction to multi-agent systems," in *Proc. 1st Int. Conf. on Multi-Agent Syst.*, San Francisco, CA, 1995.

[111]   M. Wooldridge and N. Jennings, "Intelligent agents: theory and practice," *Knowledge Eng. Review*, vol. 10, no. 2, pp. 115-152, 1995.

[112]   M. Wooldridge, J.P. Mueller, and M. Tambe, "Intelligent agents II," *Lect. Notes in Artificial Intelli.*, vol. 1037, pp. 438, 1996.

[113]   J.M. Bradshaw, "An introduction to software agents," in *Software Agents*. Cambridge, MA: MIT Press, 1997, pp. 3-46.

[114]   D.C. Dennet, *The Intentional Stance*. Cambridge, MA: MIT Press, 1987.

[115]   S. Franklin and A. Graesser, "Is it an agent or just a program? a taxonomy for autonomous agents," Springer-Verlag, London, UK, 3-540-62507-0, 1996.

[116]   T. Kaehler and D. Patterson, "A small taste of Smalltalk," *Byte*, pp. 145-159, Aug. 1986.

[117]   N. Negroponte, "Agents: from direct manipulation to delegation," in *Software Agents*, J. M. Bradshaw, Eds. Menlo Park, CA: AAAI Press, 1997.

[118]   M.P. Singh, "A theoretic framework for intentions, know-how, and communications," in *Multiagent Systems: Lect. Notes in Comput. Sci.* New York, NY: Springer-Verlag, 1994, pp. 168.

[119]   Y. Shoham, "An overview of agent-oriented programming," in *Software Agents*, J. M. Bradshaw, Eds. Menlo Park, California: AAAI Press, 1997.

[120]   O. Etzioni and D.S. Weld, "Intelligent agents on the Internet: fact, fiction, and forecast," *IEEE Expert*, vol. 10, no. 4, pp. 44-49, 1995.

[121]   D. Gilbert, M. Aparicio, B. Atkinson, S. Brady, J. Ciccarino, B. Grosof, P. O' Connor, D. Osisek, S. Pritko, R. Spagna, and L. Wilson, "IBM intelligent agent strategy," IBM, White Paper, 1995.

[122]   C. Petrie, "Agent based engineering, the Web, and intelligence," *IEEE Expert*, vol. 11, no. 6, pp. 24-29, 1996.

[123]   T.S. Perraju, "Multi-agent architectures for high assurance systems," in *Proc. Amer. Control Conf.*, 1999, pp. 3154-3157.

[124]   P. Maes, "Modeling adaptive autonomous agents," *Artificial Life J.*, vol. 1, no. 1-2, pp. 135-162, 1994.

[125]   S.J. Russell and P. Norvig, *Artificial Intelligence: A Modern Approach.* Upper Saddle River, NJ: Prentice Hall/Pearson Education, 2003.

[126]   W.S. Humphrey, *Managing the Software Process.* Reading, MA: Addison-Wesley, 1989.

[127]   I. Sommerville, *Software Engineering.* New York, NY: Addison-Wesley, 2007.

[128]   N.H. Madhavji, V. Gruhn, W. Deiters, and W. Schafer, "Prism = methodology + process-oriented environment," in *Proc. 14th Int. Conf. on Softw. Eng.*, Nice, France, 1990, pp. 277-288.

[129]   G.T. Heineman, J.E. Botsford, G. Caldiera, G.E. Kaiser, M.I. Kellner, and N.H. Madhavji, "Emerging technologies that support a software process life cycle," *IBM Syst. J.*, vol. 33, no. 3, pp. 501-529, 1994.

[130]   M.M. Tanik and E.S. Chan, *Fundamentals of Computing for Software Engineers.* Van Nostrand Reinhold, 1991.

[131]   T.W.M. Coalition, "Terminology & glossary (Issues 3.0)," Workflow Management Coalition, Winchester, UK, WFMC-TC-1011, 1999.

[132]   S. Amit, "NSF workshop on workflow and process automation in information systems: state-of-the-art and future directions," *SIGGROUP Bulletin*, vol. 18, no. 1, pp. 23-24, 1997.

[133]   W. vanderAalst and K. vanHee, *Workflow Management: Models, Methods, and Systems (Cooperative Information Systems).* Cambridge, MA: The MIT Press, 2004.

[134]   C. Soltenborn, "Analysis of UML workflow diagrams with dynamic meta modeling techniques," Diploma thesis, Dept. Comput. Sci., Univ. of Paderborn, Germany, 2006.

[135]   M. Weske, W.M.P. van der Aalst, and H.M.W. Verbeek, "Advances in business process management," *Data & Knowledge Eng.*, vol. 50, no. 1, pp. 1-8, 2004.

[136]   C. Bill, I.K. Marc, and O. Jim, "Process modeling," *Commun. ACM*, vol. 35, no. 9, pp. 75-90, 1992.

[137]   R. Conradi, C. Fernstr, A. Fuggetta, and R. Snowdon, "Towards a reference framework for process concepts," in *Proc. 2nd European Workshop on Softw. Process Technology*, 1992, pp. 3-17.

[138]   R. Yeh, R. Schlemmer, and R. Mittermeir, "A systemic approach to process modeling," *J. of Syst. Integration*, vol. 1, no. 3, pp. 265-282, 1991.

[139]   G. Bruno and R. Agarwal, "Modeling the enterprise engineering environment," *IEEE Trans. on Eng. Manage.*, vol. 44, no. 1, pp. 20-30, 1997.

[140]   M.S. Genesereth and S.P. Ketchpel, "Software agents," *Commun. ACM*, vol. 37, no. 7, pp. 48-ff, 1994.

[141]   G. Schreiber, *Knowledge Engineering and Management: The CommonKADS Methodology*. Cambridge, MA: MIT Press, 2000.

[142]   N. Glaser, *Conceptual Modelling of Multi-Agent systems: The CoMoMAS Engineering Environment*. Boston, MA: Kluwer Academic Publishers, 2002.

[143]   C.A. Iglesias, M. Garijo, J.C. González, and J.R. Velasco, "A methodological proposal for multiagent systems development extending Common-KADS," in *Proc. 10th Banff Knowledge Acquisition for Knowledge-Based Syst. Workshop*, Banff, Canada, 1996, pp. 25-37.

[144]   M. Wooldridge, N.R. Jennings, and D. Kinny, "The gaia methodology for agent-oriented analysis and design," *Autonomous Agents and Multi-Agent Syst.*, vol. 3, no. 3, pp. 285-312, 2000.

[145]   K.M. Kavi, D.C. Kung, H. Bhambhani, and G. Pandchol, "Extending UML to modeling and design of multi agent systems," in *Proc. 2nd Int. Workshop on Softw. Eng. for Large-Scale Multi-Agent Systems (SELMAS)*, Portland, OR, 2003, pp. 3–4.

[146]   G. Caire, W. Coulier, F.J. Garijo, J. Gomez, J. Pavon, F. Leal, P. Chainho, P.E. Kearney, J. Stark, R. Evans, and P. Massonet, "Agent oriented analysis using message UML," in *Proc. AOSE*, 2001, pp. 119-135.

[147]   World Wide Web Consortium. (2007). W3C semantic Web activity [Online]. Available: http://www.w3.org/2001/sw/

[148]   A.W. Brown, J. Conallen, and D. Tropeano, "Models, modeling, and model driven development," in *Model-Driven Software Development*. New York, NY: Springer, 2005, pp. 1-17.

[149]   S. Beydeda, M. Book, and V. Gruhn, *Model-Driven Software Development*. New York, NY: Springer, 2005.

[150]   J. Miller and J. Mukerji. (2003, Jun.). MDA guide. [Online]. Available: www.omg.org/docs/omg/03-06-01.pdf

[151]   A.G. Kleppe, J.B. Warmer, and W. Bast, *MDA Explained: The Model Driven Architecture: Practice and Promise*. Boston, MA: Addison-Wesley, 2003.

[152]   S.J. Mellor and M.J. Balcer, *Executable UML: A Foundation for Model-Driven Architecture*. New York, NY: Addison-Wesley, 2002.

[153]   C. Raistrick, *Model Driven Architecture with Executable UML*. Cambridge, NY: Cambridge Univ. Press, 2004.

[154]   G. Nicolas and M. Amel, "A formal framework to generate XPDL specifica-
         tions from UML activity diagrams," in *Proc. ACM Symp. on Applied Com-
         puting*, Dijon, France, 2006, pp. 1224-1231.

[155]   J. Ping, Q. Mair, and J. Newman, "Using UML to design distributed collabo-
         rative workflows: from UML to XPDL," in *Proc. 12th IEEE Int. Workshops
         on Enabling Technologies: Infrastructure for Collaborative Enterprises*,
         2003, pp. 71-76.

[156]   T.W.M. Coalition, "XML process definition language," Workflow Manage-
         ment Coalition, FL, Version 2.0, WFMC-TC-1025, 2005.

[157]   X. Ying, C. Deren, and C. Min, "Research of Web services workflow and its
         key technology based on XPDL," in *Proc. IEEE Int. Conf. on Syst., Man and
         Cybern.*, 2004, pp. 2137-2142.

[158]   W.v.d. Aalst, "Patterns and XPDL: A critical evaluation of the XML process
         definition language," Queensland Univ. of Technology, Brisbane, QUT
         Tech. Rep., FIT-TR-2003-06, 2003.

[159]   S.A. White. (2004, May). Introduction to BPMN [Online]. Available:
         http://www.bpmn.org/

[160]   BPMI.org Board of Directors. (2004, Jun.). BPMI.org phase 2: insight, inno-
         vation,   interoperability   [Online].   Available:
         www.bpmi.org/downloads/BPMI_Phase_2.pdf

[161]   World Wide Web Consortium. (2004, Dec.). Web services choreography de-
         scription   language   version   1.0   [Online].   Available:
         http://www.w3.org/TR/2004/WD-ws-cdl-10-20041217/

[162]   J. Mendling and M. Hafner, "From inter-organizational workflows to process
         execution: generating BPEL from WS-CDL," in *On the Move to Meaningful
         Internet Syst. 2005: OTM Workshops*. 2005, pp. 506-515.

[163]   C. Ouyang, W.M.P. van der Aalst, M. Dumas, terHofstede, and H.M. Arthur.
         (2006, Oct.). From business process models to process-oriented software sys-
         tems: the   BPMN   to   BPEL   way   [Online].   Available:
         http://eprints.qut.edu.au/archive/00005266/

[164]   K. Mantell. (2005, Sept.). From UML to BPEL: model driven architecture in
         a   Web   services   world   [Online].   Available:
         http://www.ibm.com/developerworks/webservices/library/ws-uml2bpel/

[165]   D. Shukla and B. Schmidt, *Essential Windows Workflow Foundation*. Upper
         Saddle River, NJ: Addison-Wesley, 2007.

[166]   Microsoft. (2007). Windows workflow foundation [Online]. Available:
         http://msdn2.microsoft.com/en-us/netframework/aa663328.aspx

[167] A. Paventhan, K. Takeda, S.J. Cox, and D.A. Nicole, "Leveraging windows workflow foundation for scientific workflows in wind tunnel applications," in *Proc. 22nd Int. Conf. on Data Eng. Workshops*, 2006, pp. 65.

[168] K. Voss, H. Genrich, G. Rozenberg, and C.A. Petri, *Concurrency and Nets: Advances in Petri Nets*. New York, NY: Springer-Verlag, 1987.

[169] T. Murata, "Petri nets: properties, analysis and applications," *J. Proc. of the IEEE*, vol. 77, no. 4, pp. 541-580, 1989.

[170] L.P. James, "Petri nets," *ACM Comput. Surv.*, vol. 9, no. 3, pp. 223-252, 1977.

[171] E. Wolfgang and G. Volker, "FUNSOFT nets: a Petri net based software process modeling language," in *Proc. 6th Int. Workshop on Softw. Specification and Design*, Como, Italy, 1991, pp. 175-184.

[172] R. Milner, J. Parrow, and D. Walker, "A calculus of mobile processes, parts {I} and {II}," *Inform. and Comput.*, vol. 100, pp. 1-40, 41-77, 1992.

[173] R. Milner, "The polyadic pi-calculus: a tutorial," in *Logic and Algebra of Specification*, F. L. Hamer, W. Brauer and H. Schwichtenberg, Eds. Springer-Verlag, 1993, pp. 203-246.

[174] J. Parrow, "An introduction to the pi-calculus," in *Handbook of Process Algebra*. New York, NY: Elsevier, 2001, pp. 1342.

[175] J.A. Bergstra, A. Ponse, and S.A. Smolka, *Handbook of Process Algebra*. New York, NY: Elsevier, 2001.

[176] R. Seker, " Component-based software modeling based on Shannon's information channels," Ph.D. dissertation, Dept. Elec. and Comput. Eng., Univ. of Alabama at Birmingham, Birmingham, AL, 2002.

[177] S.S. Muchnick and N.D. Jones, *Program Flow Analysis: Theory and Applications*. Englewood Cliffs, NJ: Prentice Hall, 1981.

[178] C.V. Ramamoorthy, "Analysis of graphs by connectivity considerations," *J. ACM*, vol. 13, no. 2, pp. 211-222, 1966.

[179] F.E. Allen, "Control flow analysis," *SIGPLAN Notes*, vol. 5, no. 7, pp. 1-19, 1970.

[180] Y. Tang, A.H. Dogru, F.J. Kurfess, and M.M. Tanik, "Computing cyclomatic complexity with cubic flowgraphs," *J. of Syst. Integration*, vol. 10, no. 4, pp. 395-409, 2001.

[181] R. Seker and M.M. Tanik, "An information theoretical framework for modeling component-based systems," *IEEE Trans. on Syst., Man, and Cybern. – Part C: Applicat. and Reviews*, no. 99, pp. 1-10, Feb. 2004.

[182]   S.F. Mills, W.F. Engel, B. Giroir, and M.M. Tanik, "A structured design methodology for process engineering," in *Proc. Int. Design and Process Technology*, Austin, TX, 1995, pp. 121-126.

[183]   P. Armenise, S. Bandinelli, C. Ghezzi, and A. Morzenti, "A survey and assessment of software process representation formalisms," *Int. J. on Softw. Eng. and Knowledge Eng.*, vol. 3, no. 3, pp. 401-426, 1993.

[184]   A.L. Opdahl and G. Sindre, "Representing real-world processes," in *Proc. 28th Hawaii Int. Conf. on Syst. Sci.*, 1995, pp. 821-830.

[185]   Stanley M. Sutton, Jr., H. Dennis, and J.O. Leon, "APPL/A: a language for software process programming," *ACM Trans. Softw. Eng. Methodol.*, vol. 4, no. 3, pp. 221-286, 1995.

[186]   S.M.J. Sutton, B.S. Lerner, and L.J. Osterweil, "Experience using the JIL process programming language to specify design processes," Univ. of Massachusetts, Cambridge, MA, UM-CS-1997-068, 1997.

[187]   J. Hong Bae and S. Hyo-Won, "The hierarchical frame of enterprise activity modeling (HF-EAM)," *IEEE Trans. on Eng. Manage.*, vol. 49, no. 4, pp. 459-478, 2002.

[188]   L. Ling and P. Calton, "A transactional activity model for organizing open-ended cooperative activities," in *Proc. 31st Hawaii Int. Conf. on Syst. Sci.*, 1998, pp. 733-743.

[189]   N. Gaducheau, E. Soulier, and M. Lewkowicz, "Design and evaluation of activity model-based groupware: methodological issues," in *Proc. 14th IEEE Int. Workshops on Enabling Technologies: Infrastructure for Collaborative Enterprise*, 2005, pp. 226-231.

[190]   A. Bajaj and S. Ram, "SEAM: A state-entity-activity-model for a well-defined workflow development methodology," *IEEE Trans. on Knowledge and Data Eng.*, vol. 14, no. 2, pp. 415-431, 2002.

[191]   D. Thomas and A. Hunt, "State machines," *IEEE Softw.*, vol. 19, no. 6, pp. 10-12, 2002.

[192]   B. Bukovics, *Pro WF: Windows Workflow in .NET 3.0 (Expert's Voice in .Net)*. Apress, 2007.

[193]   C. Rolland, C. Souveyet, and C.B. Achour, "Guiding goal modeling using scenarios," *IEEE Trans. on Softw. Eng.*, vol. 24, no. 12, pp. 1055-1071, 1998.

[194]   E. Kavakli and P. Loucopoulos, "Experiences with goal-oriented modeling of organizational change," *IEEE Trans. on Syst., Man and Cybern.*, vol. 36, no. 2, pp. 221-235, 2006.

[195] S. Zhiqi, M. Chunyan, T. Xuehong, and R. Gay, "Goal oriented modeling for intelligent software agents," in *Proc. IEEE/WIC/ACM Int. Conf. on Intell. Agent Technology*, 2004, pp. 540-543.

[196] M.A. Ould and British Computer Society, *Business Process Management: A Rigorous Approach*. Tampa, FL: Meghan-Kiffer Press, 2005.

[197] K. Cox, S. Bleistein, and J. Verner, "Connecting role activity diagrams to the problem frames approach," in *Proc. 9th Australian Workshop on Requirements Eng. (AWRE)*, Adelaide, Australia, 2004, pp. 1-13.

[198] B. Orriëns, J. Yang, and M. Papazoglou, "A framework for business rule driven Web service composition," in *Conceptual Modeling for Novel Application Domains*. 2003, pp. 52-64.

[199] S. Dustdar and W. Schreiner, "A survey on Web services composition," *Int. J. of Web and Grid Services*, vol. 1, no. 1, pp. 1-30, 2005.

[200] N. Milanovic and M. Malek, "Current solutions for Web service composition," *IEEE Internet Comput.*, vol. 8, no. 6, pp. 51-59, Nov.-Dec. 2004.

[201] J.H. Rao and X.M. Su, "A survey of automated Web service composition methods," *Semantic Web Services and Web Process Composition*, vol. 3387, pp. 43-54, 2005.

[202] K. Markus and K. Alfons, "Towards context-aware adaptable Web services," in *Proc. 13th Int. World Wide Web Conf. on Alternate Track Papers and Posters*, New York, NY, 2004, pp. 55-65.

[203] P. lvarez, J.A. Baares, P.R. Muro-Medrano, J. Nogueras, and F.J. Zarazaga, "A Java coordination tool for Web service architectures: the location-based service context," in *Proc. Revised Papers from the Int. Workshop on Scientific Eng. for Distributed Java Applicat.*, 2003, pp. 1-14.

[204] R.S. Cost, C. Ye, W.F. Timothy, L. Yannis, and P. Yun, "Using colored Petri nets for conversation modeling," in *Proc. Issues in Agent Commun.*, 2000, pp. 178-192.

[205] B. Benatallah, Q.Z. Sheng, and M. Dumas, "The self-serv environment for Web services composition," *IEEE Internet Comput.*, vol. 7, no. 1, pp. 40-48, 2003.

[206] D. Skogan, R. Gronmo, and I. Solheim, "Web service composition in UML," in *Proc. IEEE Int. Conf. of the Enterprise Distributed Object Computing*, 2004, pp. 47-57.

[207] D. Suvee, B. De Fraine, M.A. Cibran, B. Verheecke, N. Joncheere, and W. Vanderperren, "Evaluating FuseJ as a Web service composition language," in *Proc. 3rd IEEE European Conf. on Web Services*, 2005, pp. 11.

[208] P. Albert, L. Henocque, and M. Kleiner, "Configuration based workflow composition," in *Proc. IEEE Int. Conf. on Web Services*, 2005, pp. 285-292.

[209]   B. Ludscher, I. Altintas, C. Berkley, D. Higgins, E. Jaeger, M. Jones, E.A. Lee, J. Tao, and Y. Zhao, "Scientific workflow management and the Kepler system: research Articles," *Concurrent Computing: Practical Experiences*, vol. 18, no. 10, pp. 1039-1065, 2006.

[210]   V. Pankratius and W. Stucky, "A formal foundation for workflow composition, workflow view definition, and workflow normalization based on Petri nets," in *Proc. 2nd Asia-Pacific Conf. on Conceptual Modeling*, Newcastle, New South Wales, Australia, 2005, pp. 79-88.

[211]   J.P. Thomas, M. Thomas, and G. Ghinea, "Modeling of Web services flow," in *Proc. IEEE Int. Conf. on E-Commerce*, Newport Beach, CA, 2003, pp. 391-398.

[212]   A. Adi, S. Stoutenburg, and S. Tabet, *Rules and Rule Markup Languages for the Semantic Web*. New York, NY: Springer, 2005.

[213]   B. Benatallah, M. Dumas, M. Fauvet, and F. Rabhi, "Towards patterns of Web services composition," *Patterns and Skeletons for Parallel and Distributed Computing*, pp. 265-296, 2003.

[214]   C.T.H. Everaars, B. Koren, and F. Arbab, "Dynamic process composition and communication patterns in irregularly structured applications," in *Proc. 11th IPPS/SPDP Workshops held in Conjunction with the 13th Int. Parallel Processing Symp. and 10th Symp. on Parallel and Distributed Process.*, 1999, pp. 1046-1054.

[215]   C. Anis and M. Mira, "Hybrid Web service composition: business processes meet business rules," in *Proc. 2nd Int. Conf. on Service-Oriented Computing*, New York, NY, 2004, pp. 30-38.

[216]   M.T. Tut and D. Edmond, "The use of patterns in service composition," *Web Services, E-Business, and the Semantic Web*, vol. 2512, pp. 28-40, 2002.

[217]   S.M. Yacoub and H.H. Ammar, "UML support for designing software systems as a composition of design patterns," in *Proc. 4th Int. Conf. on the Unified Modeling Language, Modeling Languages, Concepts, and Tools*, 2001, pp. 149-165.

[218]   W.-L. Dong, H. Yu, and Y.-B. Zhang, "Testing BPEL-based Web service composition using high-level Petri nets," in *Proc. 10th IEEE Int. Enterprise Distributed Object Computing Conf.*, 2006, pp. 441-444.

[219]   M. Klusch, A. Gerber, and M. Schmidt, "Semantic Web service composition planning with OWLS-XPlan," in *Proc. 1st Int. AAAI Fall Symp. on Agents and the Semantic Web*, Arlington, VA, 2005, pp. 107-119.

[220]   L. Qiu, F. Lin, C. Wan, and Z. Shi, "Semantic Web services composition using AI planning of description logics," in *Proc. IEEE Asia-Pacific Conf. on Services Computing*, 2006, pp. 340.

[221]   J. Gekas and M. Fasli, "Automatic Web service composition using Web connectivity analysis," in *Proc. W3C Workshop on Frameworks for Semantics in Web Services*, Digital Enterprise Res. Inst. (DERI), Innsbruck, Austria, 2005.

[222]   K. Sivashanmugam, J.A. Miller, A.P. Sheth, and K. Verma, "Framework for semantic Web process composition," *Int. J. Electron. Commerce*, vol. 9, no. 2, pp. 71-106, 2004-2005.

[223]   T. Nie, G. Yu, D. Shen, Y. Kou, and J. Song, "An approach for composing Web services on demand," in *Proc. 10th Int. Conf. on Comput. Supported Cooperative Work in Design (CSCWD)*, 2006, pp. 1-6.

[224]   T. Morita, Y. Hirano, Y. Sumi, S. Kajita, and K. Mase, "A pattern mining method for interpretation of interaction," in *Proc. 7th Int. Conf. on Multimodal Interfaces*, Toronto, Italy, 2005, pp. 267-273.

[225]   W.T. Tsai, H. Qian, S. Xin, and C. Yinong, "Dynamic collaboration simulation in service-oriented computing paradigm," in *Proc. 40th Annu. Simulation Symp.*, 2007, pp. 41-48.

[226]   S. Lammermann and E. Tyugu, "A specification logic for dynamic composition of services," in *Proc. Int. Conf. on Distributed Computing Syst. Workshop*, 2001, pp. 157-162.

[227]   D. Thakker, T. Osman, and D. Al-Dabass, "Web services hybrid dynamic composition models for ubiquitous computing networks," in *Proc. 8th Int. Conf. on Advanced Commun. Technology*, 2006, pp. 7.

[228]   Z. Chen, J. Ma, L. Song, and L. Lian, "An efficient approach to Web services discovery and composition when large scale services are available," in *Proc. IEEE Asia-Pacific Conf. on Services Computing (APSCC)*, 2006, pp. 34.

[229]   R.E. Filman, *Aspect Oriented Software Development*. UK: Addison-Wesley, 2005.

[230]   C. Alexander, *The Timeless Way of Building*. New York, NY: Oxford Univ. Press, 1979.

[231]   N. Francez, C.A.R. Hoare, D.J. Lehmann, and W.P. de Roever, *Semantics of Nondeterminism, Concurrency and Communication*. Utrecht, 1978.

[232]   E.W. Dijkstra, *A Discipline of Programming*. Englewood Cliffs, NJ: Prentice Hall, 1976.

[233]   Wikipedia-Contributors. (2007, Sept.). Unbounded nondeterminism [Online]. Available: http://en.wikipedia.org/w/index.php?title=Unbounded_nondeterminism&oldid=122540375

[234]   C.A.R. Hoare, *Communicating Sequential Processes*. Englewood Cliffs, NJ: Prentice/Hall International, 1985.

[235]   B.D. Janz, J.A. Colquitt, and R.A. Noe, "Knowledge worker team effectiveness: the role of autonomy, interdependence, team development, and contextual support variables," *Personnel Psychology*, vol. 50, no. 4, pp. 877-904, 1997.

[236]   Y. Ye, Y. Yamamoto, and K. Kishida, "Dynamic community: a new conceptual framework for supporting knowledge collaboration in software development," in *Proc. 11th Asia-Pacific Softw. Eng. Conf. (APSEC)*, 2004, pp. 30-37.

[237]   G. Kratkiewicz and G. Mitchell, "An adaptive semantic approach to personal information management," in *Proc. IEEE Int. Conf. on Syst., Man and Cybern.*, 2004, pp. 1395-1400.

[238]   M.G.D. Paula, S.D.J. Barbosa, C. Jos, and P.d. Lucena, "Conveying human-computer interaction concerns to software engineers through an interaction model," in *Proc. Latin Amer. Conf. on Human-Comput. Interaction*, Cuernavaca, Mexico, 2005, pp. 109-119.

[239]   R.R. Penner, K.S. Nelson, and N.H. Soken, "Facilitating human interactions in mixed initiative systems through dynamic interaction generation," in *Proc. IEEE Int. Conf. on Syst., Man, and Cybern. (Computational Cybern. and Simulation)*, 1997, pp. 714-719.

[240]   C. Sedogbo and Human Interaction Technologies Lab, "Human-system interaction container paradigm," in *Proc. W3C Workshop on Multimodal Interaction*, Sophia Antipolis, France, 2004.

[241]   S. Voida, E.D. Mynatt, B. MacIntyre, and G.M. Corso, "Integrating virtual and physical context to support knowledge workers," *IEEE Pervasive Comput.*, vol. 1, no. 3, pp. 73-79, 2002.

[242]   X. Kong, L. Liu, and D. Lowe, "Separation of concerns: a Web application architecture framework," *J. of Digital Inform.*, vol. 6, no. 2, 2005.

[243]   Chief Information Officer Council. (2001, Feb.). A practical guide for developing an enterprise architecture. [Online]. Available: www.gao.gov/bestpractices/bpeaguide.pdf

[244]   T.O. Group. (2007). Welcome to TOGAF™ - the Open Group architecture framework [Online]. Available: http://www.opengroup.org/architecture/togaf8-doc/arch/

[245]   K. Harmon, "The "systems" nature of enterprise architecture," in *Proc. IEEE Int. Conf. on Syst., Man and Cybern.*, 2005, pp. 78-85.

[246]   D.D. Villiers. (2003, Dec.). Using the Zachman framework to assess the rational unified process. [Online]. Available: http://www-128.ibm.com/developerworks/rational/library/372.html

[247]   J.A. Zachman, "A framework for information systems architecture," *IBM Syst. J.*, vol. 26, no. 3, pp. 276-292, 1987.

[248] J. Schekkerman, *The Economic Benefits of Enterprise Architecture*. Trafford Publishing, 2005.

[249] P.G. Carlock and R.E. Fenton, "System of Systems (SoS) enterprise systems engineering for information-intensive organizations," *Syst. Eng.*, vol. 4, no. 4, pp. 242-261, 2001.

[250] P. Carlock and J.A. Lane, "System of systems enterprise systems engineering, the enterprise architecture management framework, and system of systems cost estimation," in *Proc. 21st Int. Forum on COCOMO and Softw. Cost Modeling*, Herndon, VA, 2006.

[251] The Open Group, *The Open Group Architecture Framework*. The Open Group, 2006.

[252] K.R. Conner and C.K. Prahalad, "A resource-based theory of the firm: knowledge versus opportunism," *Org. Sci.*, vol. 7, no. 5, pp. 477-501, 1996.

[253] H. Demsetz, "The theory of the firm revisited," in *The Nature of the Firm: Origins, Evolution, and Development*, Williamson and S. G. Winter, Eds. New York, NY: Oxford Univ. Press, 1993, pp. 159-178.

[254] R.M. Grant, "Toward a knowledge-based theory of the firm," *Strategic Manage. J.*, vol. 17, Winter Special, pp. 109-122, 1996.

[255] B. Kogut and U. Zander, "What firms do? Coordination, identity, and learning," *Org. Sci.*, vol. 7, no. 5, pp. 502-518, 1996.

[256] K. Kosanke, "CIMOSA - overview and status," *Comput. Ind.*, vol. 27, no. 2, pp. 101-109, Oct. 1995.

[257] D.C. Galunic and S. Rodan, "Resource recombinations in the firm: knowledge structures and the potential for schumpeterian innovation," *Strategic Manage. J.*, vol. 19, no. 12, pp. 1193-1201, 1998.

[258] D.J. Teece, G. Pisano, and A. Shuen, "Dynamic capabilities and strategic management," *Strategic Manage. J.*, vol. 18, no. 7, pp. 509-533, 1997.

[259] M. Polanyi, *The Tacit Dimension*. Gloucester, MA: Peter Smith, 1983.

[260] I. Nonaka and H. Takeuchi, *The Knowledge-Creating Company: How Japanese Companies Create the Dynamics of Innovation*. New York, NY: Oxford Univ. Press, 1995.

[261] R. Nelson and S. Winter, *Evolutionary Theory of Economic Change*. MA: Belknap Press, 1985.

[262] K.E. Weick and K.H. Roberts, "Collective mind in organizations: heedful interrelating on flight decks," *Administ. Sc. Quart.*, vol. 38, no. 3, pp. 357-381, 1993.

[263]   J. Martin and J.J. Odell, *Object-Oriented Methods: A Foundation*. Upper Saddle River, NJ: Prentice Hall, 1998.

[264]   T. Morgan, *Business Rules and Information Systems: Aligning IT with Business Goals*. Boston, MA: Addison-Wesley, 2002.

[265]   R.S. Sandhu, E.J. Coyne, H.L. Feinstein, and C.E. Youman, "Role-based access control models," *IEEE Comput.*, vol. 29, no. 2, pp. 38-47, 1996.

[266]   D.E. O'Leary, "Enterprise knowledge management," *IEEE Comput.*, vol. 31, no. 3, pp. 54-61, 1998.

[267]   D. Fensel, *Ontologies: A Silver Bullet for Knowledge Management and Electronic Commerce*. New York, NY: Springer-Verlag, 2003.

[268]   A.F. Tuzovsky, V.Z. Yampolsky, and S.V. Chirikov, "Ontological knowledge management system development," in *Proc. The 9th Russian-Korean Int. Symp. on Sci. and Technology*, 2005, pp. 718-720.

[269]   E. Friedman-Hill, *JESS in Action*. Greenwich, CT: Manning Publications Co., 2003.

[270]   J.M. Kabasele Tenday, J.J. Quisquater, and M. Lobelle, "Deriving a role-based access control model from the OBBAC model," in *Proc. IEEE 8th Int. Workshops on Enabling Technologies: Infrastructure for Collaborative Enterprises*, 1999, pp. 147-151.

[271]   V. Cridlig, R. State, and O. Festor, "An integrated security framework for XML based management," in *Proc. 9th IFIP/IEEE Int. Symp. on Integrated Network Manage.*, 2005, pp. 587-600.

[272]   D. Ferraiolo, D.R. Kuhn, and R. Chandramouli, *Role-Based Access Control*. Boston, MA: Artech House, 2003.

[273]   D.F. Ferraiolo, R. Kuhn, R. Chandramouli, and J. Barkley. (2007). NIST role based access control [Online]. Available: http://csrc.nist.gov/rbac/

[274]   D. Brickley and L. Miller. (2000). Friend of a friend (FOAF) project [Online]. Available: http://www.foaf-project.org/

[275]   X. Jiang and A.-H. Tan, "Learning and inferencing in user ontology for personalized semantic Web services," in *Proc. 15th Int. Conf. on World Wide Web*, Edinburgh, Scotland, 2006, pp. 1067-1068.

[276]   M. Pazzani and D. Billsus, "Learning and revising user profiles: the identification of interesting Web sites," *Mach. Learning*, vol. 27, no. 3, pp. 313-331, 1997.

[277]   E. Van der Vlist, *XML Schema*. Sebastopol, CA: O'Reilly, 2002.

[278]   A.V. Aho and J.D. Ullman, "Optimal partial-match retrieval when fields are independently specified," *ACM Trans. Database Syst.*, vol. 4, no. 2, pp. 168-179, 1979.

[279]   S. Dustdar and T. Hoffmann, "Interaction pattern detection in process oriented information systems," *Data & Knowledge Eng.*, vol. 62, no. 1, pp. 138-155, 2007.

[280]   S. Thompson, T. Torabi, and P. Joshi, "A framework to detect deviations during process enactment," in *Proc. 6th IEEE/ACIS Int. Conf. on Comput. and Inform. Sci.*, 2007, pp. 1066-1073.

[281]   D.E. Perry and A.L. Wolf, "Session 1: people, processes, and practice," in *Proc. 9th Int. Soft. Process Workshop*, 1994, pp. 2.

[282]   M. Dai, P. Wang, A.D. Boyd, G. Kostov, B. Athey, E.G. Jones, W.E. Bunney, R.M. Myers, T.P. Speed, H. Akil, S.J. Watson, and F. Meng, "Evolving gene/transcript definitions significantly alter the interpretation of GeneChip data," *Nucleic Acids Res.*, vol. 33, no. 20, pp. 175, 2005.

[283]   W. Dosch, "A loose interaction pattern for asynchronous components," in *Proc. 12th Asia-Pacific Softw. Eng. Conf. (APSEC)*, 2005, pp. 9.

[284]   A. Barros, M. Dumas, and A.H.M. ter Hofstede, "Service interaction patterns," in *Business Process Management*. 2005, pp. 302-318.

[285]   F. McElroy. (2003, Jun.). Architecture patterns for solution integration for websphere business integration [Online]. Available: http://www.ibm.com/developerworks/websphere/library/techarticles/0306_mcelroy/mcelroy.html

[286]   S. Wasserman and K. Faust, *Social Network Analysis: Methods and Applications*. Cambridge, NY: Cambridge Univ. Press, 1994.

[287]   S.C. Thompson, "Process of selecting message trunk configurations for an overlay packet voice tandem," M.S. thesis, Dept. Elec.and Comput. Eng., Univ. of Alabama at Birmingham, Birmingham, AL, 2004.

[288]   S. Koruna, "Leveraging knowledge assets: combinative capabilities - theory and practice," *R&D Manage.*, vol. 34, no. 5, pp. 505-516, 2004.

[289]   G. Hamel and C.K. Prahalad, "The core competence of the corporation," *Harvard Bus. Review*, vol. 68, no. 3, pp. 12, 1990.

[290]   K.P. Coyne, S.J.D. Hall, and P.G. Clifford, "Is your core competence a mirage," *The McKinsey Quart.*, no. 1, pp. 40-54, 1997.

[291]   A. Osterwalder, Y. Pigneur, and C.L. Tucci, "Clarifying business models: origins, present, and future of the concept," *Commun. of the Assoc. for Inform. Syst.*, vol. 16, pp. 1-25, 2005.

[292]   J. Gordijn, H. Akkermans, and H. van Vliet, "Business modeling is not process modeling," in *Conceptual Modeling for E-Business and the Web*. Springer-Verlag, 2000, pp. 40.

[293]   A. Osterwalder and Y. Pigneur. (2002, Feb). An e-business model ontology for modeling e-business. [Online]. Available: http://ideas.repec.org/p/wpa/wuwpio/0202004.html

[294]   M. Dubinko, *XForms Essentials*. Farnham: O'Reilly, 2003.

[295]   H. Thomas and M. Pedro, "VVT terminology: a proposal," *IEEE Expert: Intell. Syst. and their Applicat.*, vol. 8, no. 3, pp. 48-55, 1993.

[296]   J.O. Grady, *System Validation and Verification*. Boca Raton, FL: CRC Press, 1998.

[297]   E.C. Jonathan and L.W. Alexander, "Software process validation: quantitatively measuring the correspondence of a process to a model," *ACM Trans. Softw. Eng. Methodol.*, vol. 8, no. 2, pp. 147-176, 1999.

[298]   S.C. Thompson and M.M. Tanik, "Analysis of large scale component based systems," in *Proc. Int. Design and Process Technology*, San Diego, CA, 2006, pp. 360-368.

[299]   S.N. Twigger, M. Shimoyama, S. Bromberg, A.E. Kwitek, and H.J. Jacob, "The Rat Genome Database, update 2007 – easing the path from disease to data and back again," *Nucleic Acids Res.*, vol. 35, database issue, pp. 658-662, Jan. 2007.

[300]   J.A. Blake, J.E. Richardson, C.J. Bult, J.A. Kadin, and J.T. Eppig, "The Mouse Genome Database (MGD): the model organism database for the laboratory mouse," *Nucleic Acids Res.*, vol. 30, no. 1, pp. 113-115, Jan. 1, 2002.

APPENDIX A

THE WCS BPEL PROCESS WITHOUT APPROVAL TASK

This appendix contains the code and screen shots of the implemented WCS BPEL process without approval task of the user. Table A.1 shows the outline of this appendix.

TABLE A.1
OUTLINE OF APPENDIX

Item	Explanation
1	BPEL process of the WCS without approval task.
2	Screen shot of the user form of the implemented process of the WCS without approval task.
3	Screen shot of the audit trail of the implemented process of the WCS without approval task.

1. BPEL Process of the WCS without Approval Task

Fig A.1 depicts the BPEL process of the WCS implemented in oracle BPEL designer.

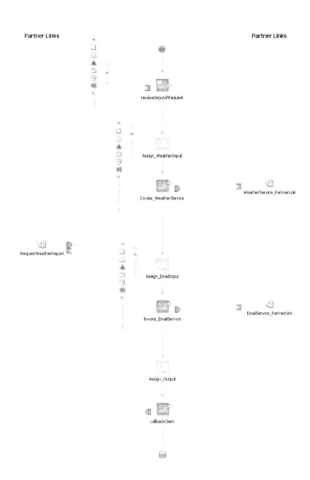

Fig. A.1. WCS implemented in Oracle BPEL designer.

2. Screen shot of the User Form of theIimplemented Process of the WCS Without Approval Task

Fig A.2 shows the screen shot of the user form of the WCS.

Fig. A.2. User form of the WCS.

3. *Screen shot of the audit trail of the implemented process of the WCS without approval task*

Fig A.3 shows the screen shot of the audit trail of the WCS.

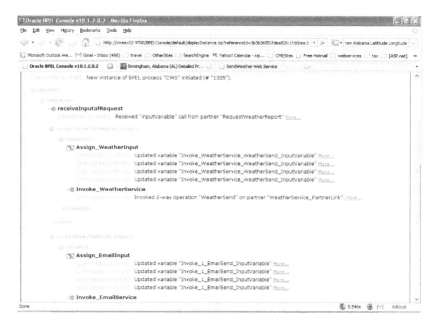

Fig. A.3. Audit trail of the WCS.

APPENDIX B

WEB-BASED SYSTEM FOR CONANT DECOMPOSITON APPROACH

This appendix contains the screen shots of the implemented interaction pattern mining approach using Conant's decomposition approach. Table B.1 shows the outline of this appendix.

TABLE B.1
OUTLINE OF APPENDIX

Item	Explanation
1	Interface to input values for random simulation of a system.
2	Computed transmission values for the generated system.
3	Simulated data of the system.
4	Transmission results that can be automatically viewed in Microsoft Excel.
5	Graphical representation of the interactions between sub-systems.
6	Code to calculate the transmission values.

1. Interface to Input Values for Random Simulation of a System

Fig B.1 depicts the Web-based interface to input values for random simulation of a system.

Generate A System

Home > Conant Intro > Generate a System

Generate a System to Test

System Name   wcs

No of Components of the System   4 ⌄

Please keep the number of times instances less than 1000

No of Time Instances Recorded   1000

No of State Types   3 ⌄

CONTINUE ▶

◀ BACK

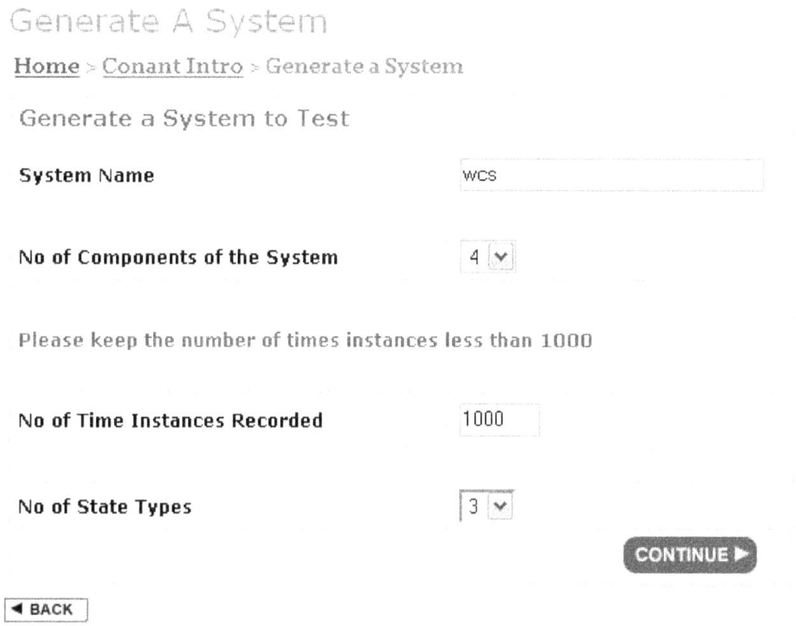

Fig B.1. Interface for random simulation of a system.

2. Computed Transmission Values for Generated System

Fig B.2 depicts the interface to compute normalized transmission values.

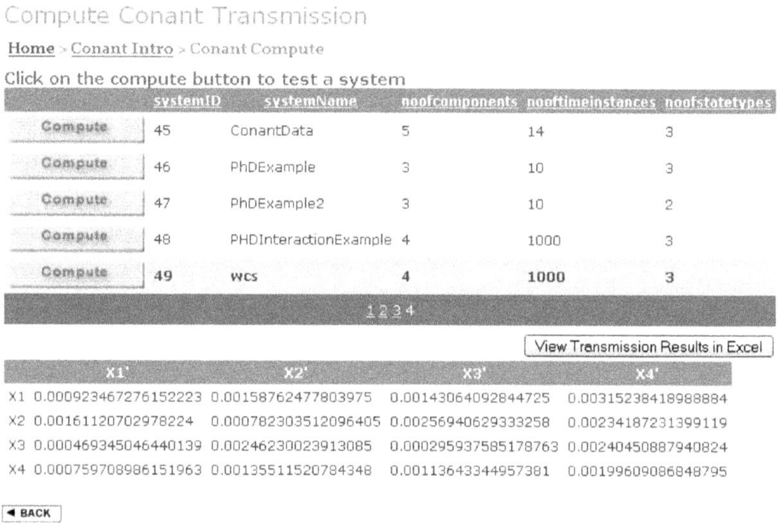

Fig. B.2. Interface to compute normalized transmission values.

3. Simulated Data of the System

Fig B.3 shows the simulated data of the system.

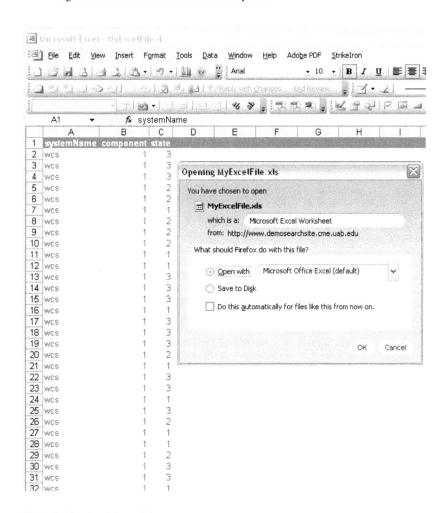

Fig. B.3. Simulated data of the system.

4. Transmission Results that can be Automatically Viewed in Microsoft Excel

Fig B.4 shows the computed normalized transmission values in Microsoft Excel.

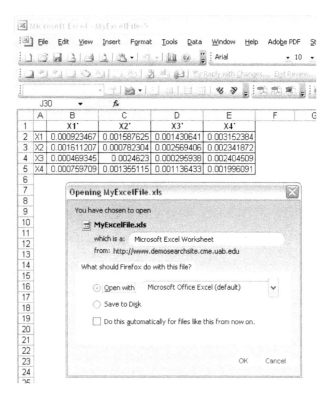

Fig. B.4. Computed normalized transmission values in Microsoft Excel.

.

5. Graphical Representation of the Interactions Between Sub-Systems.

Fig B.5 shows the graphical representation of the interactions between sub-systems.

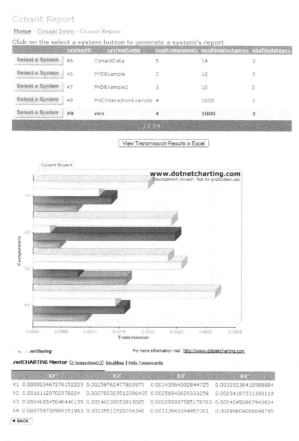

Fig. B.5. Graphical representation of the interactions between sub-systems.

6. Code to Calculate the Transmission Values

The code to calculate the normalized transmission is given below.

```
using System;
using System.Collections.Generic;
using System.Text;
using System.IO;
using System.Xml;
```

166

```
namespace ConantApplication
{
    public class ConantCompute
    {
        private ConantDataAccess da;
        private ConantDataDetails dd;

        private float[] currentstate;
        private float[,] transitionstate;

        private float NoOfObservations = 0;

        //values for computation
        private float comp = 0, comp1 = 0;
        public ConantCompute()
        {
        }
        public ConantCompute(int systemidentifier)
        {
            da = new ConantDataAccess();
            //Obtaining the value of input matrix, no of components,
no of states from DB
            dd = da.ObtainConantData(systemidentifier);
            currentstate = new float[dd.Noofstatetypes];
            transitionstate    =    new    float[dd.Noofstatetypes,
dd.Noofstatetypes];
        }
        private float CalculateLogtoBase2(float x)
        {
            float logval;
            if (x == 0)
                logval = 0;
            else
                logval = (float)Math.Log(x, 2);
            return logval;
        }
        private void ResetAllValues()
        {
            for (int k = 0; k < dd.Noofstatetypes; k++)
            {
```

```
            currentstate[k] = 0;
        }
        for (int k = 0; k < dd.Noofstatetypes; k++)
        {
            for (int l = 0; l < dd.Noofstatetypes; l++)
            {
                transitionstate[k, l] = 0;
            }
        }
        NoOfObservations = 0;
        //values for computation
        comp = 0;
        comp1 = 0;
    }
    #region calculateConantTransmission()

    //States can only start from 1
    public void CalculateConantTransmission()
    {
        //Initializing some values used in the program
        float[] hMatrix = new float[dd.Noofcomponents];
        float[] hprimeMatrix = new float[dd.Noofcomponents];
        float hjoin = 0;
        float transmission = 0;

        //I am getting the input from a DB.
        int[,] inputMatrix = dd.Inputmatrixarray;


        #region Computing H and H prime matrix
        //Calculate H and H prime Matrix
        for (int i = 0; i < dd.Noofcomponents; i++)
        {

            #region hmatrix
            //H Matrix
            ResetAllValues();

            //for loop is from 0 - (no of states -1)
            for (int j = 0; j < dd.Nooftimeinstances - 1; j++)
```

```
                {
                    for (int k = 0; k < dd.Noofstatetypes; k++)
                    {
                        if (inputMatrix[i, j] == k + 1)
                            currentstate[k]++;

                    }
                }

                for (int k = 0; k < dd.Noofstatetypes; k++)
                {
                    NoOfObservations    =    NoOfObservations    +
currentstate[k];

                }
                comp = CalculateLogtoBase2(NoOfObservations);

                for (int k = 0; k < dd.Noofstatetypes; k++)
                {
                    comp1    =    comp1    +    (currentstate[k]    *
CalculateLogtoBase2(currentstate[k]));

                }
                hMatrix[i] = comp - comp1 / NoOfObservations;
                #endregion

                #region hprime
                //Hprime
                //for loop is from 0 - (no of states -1)
                //setting all values = 0
                ResetAllValues();
                for (int k = 0; k < dd.Noofstatetypes; k++)
                {
                    currentstate[k] = 0;
                }

                for (int j = 1; j < dd.Nooftimeinstances; j++)
                {
                    for (int k = 0; k < dd.Noofstatetypes; k++)
                    {
                        if (inputMatrix[i, j] == k + 1)
```

169

```
                                    currentstate[k]++;

                    }
                }

                for (int k = 0; k < dd.Noofstatetypes; k++)
                {
                    NoOfObservations      =      NoOfObservations      +
currentstate[k];

                }

                comp = CalculateLogtoBase2(NoOfObservations);

                for (int k = 0; k < dd.Noofstatetypes; k++)
                {
                    comp1     =     comp1     +     (currentstate[k]     *
CalculateLogtoBase2(currentstate[k]));
                }

                hprimeMatrix[i] = comp - comp1 / NoOfObservations;
                #endregion
            }
            #endregion

            #region Joint Matrix and Trans
            //Calculate Joint Matrix

            for (int h = 0; h < dd.Noofcomponents; h++)
            {
                for (int i = 0; i < dd.Noofcomponents; i++)
                {
                    ResetAllValues();

                    for (int j = 0; j < dd.Nooftimeinstances - 1;
j++)
                    {
                        for (int k = 0; k < dd.Noofstatetypes; k++)
                        {
                            for (int l = 0; l < dd.Noofstatetypes;
l++)
```

```
                              {
                                  if ((inputMatrix[h, j] == k + 1) &&
(inputMatrix[i, j + 1] == l + 1))
                                      transitionstate[k, l]++;
                              }
                          }
                      }

                 for (int k = 0; k < dd.Noofstatetypes; k++)
                 {
                     for (int l = 0; l < dd.Noofstatetypes; l++)
                     {
                          NoOfObservations   =   NoOfObservations   +
transitionstate[k, l];
                     }
                 }

                 comp = CalculateLogtoBase2(NoOfObservations);

                 for (int k = 0; k < dd.Noofstatetypes; k++)
                 {
                     for (int l = 0; l < dd.Noofstatetypes; l++)
                     {

                          comp1 = comp1 + (transitionstate[k, l] *
CalculateLogtoBase2(transitionstate[k, l]));

                     }
                 }

                 hjoin = comp - comp1 / NoOfObservations;

                 transmission = hMatrix[h] + hprimeMatrix[i] -
hjoin;

                 if (hprimeMatrix[i] <= 0)
                     transmission = 0;
                 else
                     transmission      =      transmission      /
hprimeMatrix[i];
```

```
                    //h+1 to prevent component being referred to as 0
in the DB
                    string component = "X" + (h + 1);
                    string componentprime = "X" + (i + 1) + "p";

                    //Adding values to the DB
                    da.AddConantTransmissionData(transmission,
component, componentprime, TestSystemIdentifier);

                }
            }
            #endregion

        }//End calculateConantTransmission()
        #endregion

    }//End Class ComputeConant
}//End namespace ConantApplication
```

APPENDIX C

RULE DESCRIPTION FOR BOOK RECOMMENDATION AGENT

A simple recommendation agent developed using JESS is given below [269].

```
(deftemplate recommend (slot name)(slot price)
)

(defquery find-the-user-books
"Find books in price range"
(declare (variables ?key ?lower ?upper))
(book (keyword ?key)(price ?p&:(and (> ?p ?lower)(< ?p ?upper))))
)
(defquery all-books
"Get all books"
(book)
)
(defrule recommend-book
"recommend the book"
(declare (auto-focus TRUE))
(book (name ?name)(price ?price)(keyword ?key))
(not (book (price ?p&:(< ?p ?price))))
=>
(assert(recommend (name ?name)(price ?price)))
)

(defrule sort-print
?r1 <- (recommend (name ?name)(price ?price))
=>
(store priceOUT ?price)
(store nameOUT ?name)
(retract ?r1)
)
```

APPENDIX D

THE WCS BPEL PROCESS WITH APPROVAL TASK

This appendix contains the code and screen shots of the implemented WCS BPEL process without approval task of the user. Table D.1 provides the list of items contained in this appendix.

TABLE D.1
OUTLINE OF APPENDIX

Item	Explanation
1	BPMN model of the WCS with approval task.
2	Auto-generated BPEL Process of the WCS with approval task.
3	Screen shot of the user form of the implemented process of the WCS without approval task – Input for the Weather Service.
4	Screen shot of the administration page of the implemented process of the WCS with approval task – Shows that a task is on the employee queue.
5	Screen shot of the administration page of the implemented process of the WCS with approval task – Approval form.
6	Screen shot of the user form of the implemented process of the WCS without approval task – Email received by user.

1. BPMN model of the WCS with approval task

Fig. F.1 shows the BPMN model of the WCS implemented in Intalio BPMN designer.

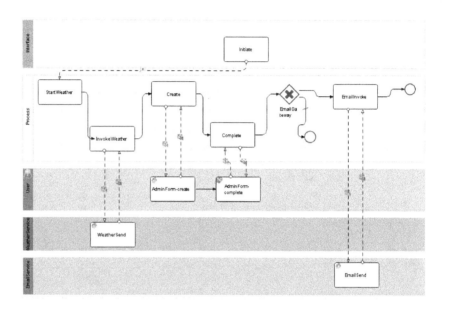

Fig. F.1. BPMN model of the WCS implemented in Intalio BPMN designer.

2. Auto-Generated BPEL Process of the WCS With Approval Task

The auto generated BPEL code of the WCS is provided below. The code below shows the inclusion of features designed to incorporate CPP semantics, such as roles of the user, Web form for the user task, and rules for the task.

```
<!--

//////////////////////////////////////////////////////////////////////
/////////////////////////
  // Author:  rsadasivam
  // Purpose: Asynchronous BPEL Process

//////////////////////////////////////////////////////////////////////
/////////////////////////
-->
<?xml version='1.0' encoding='UTF-8'?>
<bpel:process
xmlns:bpel="http://schemas.xmlsoap.org/ws/2004/03/business-process/"
xmlns:pnlk="http://schemas.xmlsoap.org/ws/2004/03/partner-link/"
xmlns:xs="http://www.w3.org/2001/XMLSchema"
xmlns:wsdl="http://schemas.xmlsoap.org/wsdl/"
xmlns:this="http://example.com/CompositeProcess/XWS/Process"
xmlns:EmailService="http://example.com/CompositeProcess/XWS/EmailServ
ice"            xmlns:tns="http://www.example.org/SendWeatherSchema"
xmlns:Interface="http://example.com/CompositeProcess/XWS/Interface"
xmlns:SendWeather="http://10.30.40.30/"
xmlns:WeatherService="http://example.com/CompositeProcess/XWS/Weather
Service"         xmlns:diag="http://example.com/CompositeProcess/XWS"
xmlns:User="http://example.com/CompositeProcess/XWS/User"
xmlns:EmailSchema="http://www.example.org/EmailSchema"
xmlns:xform="http://example.com/AdminForm/AdminForm/xform"
xmlns:bpmn="http://www.intalio.com/bpms"                    queryLan-
guage="urn:oasis:names:tc:wsbpel:2.0:sublang:xpath2.0" expressionLan-
guage="urn:oasis:names:tc:wsbpel:2.0:sublang:xpath2.0"
bpmn:label="Process" bpmn:id="_UHc1sGzBEdyvXYJHfujmvg" name="Process"
targetNamespace="http://example.com/CompositeProcess/XWS/Process">
  <bpel:import         namespace="http://10.30.40.30/"         loca-
tion="../Services/EmailService/SendEmail.wsdl"               import-
Type="http://schemas.xmlsoap.org/wsdl/" />
```

```
  <bpel:import          namespace="http://10.30.40.30/"          loca-
tion="../Services/WeatherService/SendWeather.wsdl"          import-
Type="http://schemas.xmlsoap.org/wsdl/" />
  <bpel:import                                              name-
space="http://example.com/AdminForm/AdminForm/xform"          loca-
tion="../AdminForm/AdminForm.xform.all.wsdl"              import-
Type="http://schemas.xmlsoap.org/wsdl/" />
  <bpel:import    namespace="http://example.com/CompositeProcess/XWS"
location="XWS.wsdl" importType="http://schemas.xmlsoap.org/wsdl/" />
  <bpel:import                                              name-
space="http://example.com/CompositeProcess/XWS/Process"          loca-
tion="XWS-Process.wsdl" importType="http://schemas.xmlsoap.org/wsdl/"
/>
  <bpel:partnerLinks>
    <bpel:partnerLink  name="processAndInterfacePlkVar"  partnerLink-
Type="diag:ProcessAndInterface" myRole="Process_for_Interface" />
    <bpel:partnerLink
name="processAndWeatherServiceForPortSendWeatherSoapPlkVar"  partner-
LinkType="diag:ProcessAndWeatherServiceForPortSendWeatherSoapPlk"
initializePartnerRole="true" partnerRole="WeatherService_for_Process"
/>
    <bpel:partnerLink     name="wFmagic_v0TFUGzDEdy8_fgUbrSKpgPlkVar"
partnerLinkType="diag:WFmagic_v0TFUGzDEdy8_fgUbrSKpg" initializePart-
nerRole="false"   myRole="Process_for_UserForThePortTypexformProcess"
partnerRole="User_for_ProcessForXformPort" />
    <bpel:partnerLink
name="emailServiceAndProcessForPortSendEmailSoapPlkVar"  partnerLink-
Type="diag:EmailServiceAndProcessForPortSendEmailSoapPlk" initialize-
PartnerRole="true" partnerRole="EmailService_for_Process" />
  </bpel:partnerLinks>
  <bpel:variables>
    <bpel:variable      name="thisStartWeatherRequestMsg"      mes-
sageType="this:StartWeatherRequest" />
    <bpel:variable     name="sendWeatherWeatherSendRequestMsg"    mes-
sageType="SendWeather:WeatherSendSoapIn" />
    <bpel:variable    name="sendWeatherWeatherSendResponseMsg"    mes-
sageType="SendWeather:WeatherSendSoapOut" />
    <bpel:variable       name="xformCreateTaskRequestMsg"       mes-
sageType="xform:createTaskRequest" />
    <bpel:variable       name="xformCreateTaskResponseMsg"       mes-
sageType="xform:createTaskResponse" />
```

```
    <bpel:variable   name="xformNotifyTaskCompletionRequestMsg"   mes-
sageType="xform:notifyTaskCompletionRequest" />
    <bpel:variable   name="xformNotifyTaskCompletionResponseMsg"   mes-
sageType="xform:notifyTaskCompletionResponse" />
    <bpel:variable     name="sendWeatherEmailSendRequestMsg"     mes-
sageType="SendWeather:EmailSendSoapIn" />
    <bpel:variable     name="sendWeatherEmailSendResponseMsg"     mes-
sageType="SendWeather:EmailSendSoapOut" />
  </bpel:variables>
  <bpel:sequence>
    <bpel:receive   partnerLink="processAndInterfacePlkVar"   port-
Type="this:ForInterface"          operation="StartWeather"          vari-
able="thisStartWeatherRequestMsg"              createInstance="yes"
bpmn:label="StartWeather" bpmn:id="_fKAc0GzEEdy8_fgUbrSKpg" />
    <bpel:assign name="init-variables-Process">
      <bpel:copy        bpmn:label="$sendWeatherWeatherSendRequestMsg
out:_vkFpoGzEEdy8_fgUbrSKpg">
        <bpel:from>
          <bpel:literal>
<SendWeather:WeatherSend>
  <SendWeather:request>
    <SendWeather:Latitude>
    </SendWeather:Latitude>
    <SendWeather:Longitude>
    </SendWeather:Longitude>
    <SendWeather:Startdate>
    </SendWeather:Startdate>
    <SendWeather:Numdays>
    </SendWeather:Numdays>
  </SendWeather:request>
</SendWeather:WeatherSend>
          </bpel:literal>
        </bpel:from>

<bpel:to>$sendWeatherWeatherSendRequestMsg.parameters</bpel:to>
      </bpel:copy>
      <bpel:copy              bpmn:label="$xformCreateTaskRequestMsg
out:_z9IGcGzEEdy8_fgUbrSKpg">
        <bpel:from>
          <bpel:literal>
<xform:createTaskRequest>
```

```xml
<xform:taskMetaData>
  <xform:taskId>_v0TFUGzDEdy8_fgUbrSKpg</xform:taskId>
  <xform:taskState>
  </xform:taskState>
  <xform:taskType>
  </xform:taskType>
  <xform:description />
  <xform:processId>
  </xform:processId>
  <xform:creationDate>
  </xform:creationDate>
  <xform:userOwner />
  <xform:roleOwner>examples\employee</xform:roleOwner>
  <xform:claimAction>
    <xform:user>
    </xform:user>
    <xform:role>
    </xform:role>
  </xform:claimAction>
  <xform:revokeAction>
    <xform:user>
    </xform:user>
    <xform:role>
    </xform:role>
  </xform:revokeAction>
  <xform:saveAction>
    <xform:user>
    </xform:user>
    <xform:role>
    </xform:role>
  </xform:saveAction>
  <xform:completeAction>
    <xform:user>
    </xform:user>
    <xform:role>
    </xform:role>
  </xform:completeAction>

<xform:formUrl>oxf://PhDDemoCWS/AdminForm/AdminForm.xform</xform:formUrl>
  <xform:failureCode>
```

```
    </xform:failureCode>
    <xform:failureReason>
    </xform:failureReason>

<xform:userProcessCompleteSOAPAction>http://example.com/AdminForm/Adm
in-
Form/xform/Process/notifyTaskCompletion</xform:userProcessCompleteSOA
PAction>
    <xform:isChainedBefore>
    </xform:isChainedBefore>
    <xform:previousTaskId>
    </xform:previousTaskId>

<xform:userProcessEndpoint>http://localhost:8080/ode/processes/PhDDem
oCWS/CompositeProcess/XWS/Process/User/Process_for_UserForThePortType
xformProcessPort</xform:userProcessEndpoint>

<xform:userProcessNamespaceURI>http://example.com/AdminForm/AdminForm
/xform</xform:userProcessNamespaceURI>
  </xform:taskMetaData>
  <xform:participantToken>
  </xform:participantToken>
  <xform:taskInput>
    <xform:input>
      <xform:weatherreport>
      </xform:weatherreport>
    </xform:input>
  </xform:taskInput>
</xform:createTaskRequest>
          </bpel:literal>
        </bpel:from>
        <bpel:to>$xformCreateTaskRequestMsg.root</bpel:to>
      </bpel:copy>
      <bpel:copy    bpmn:label="$xformNotifyTaskCompletionResponseMsg
out:_00QiwGzEEdy8_fgUbrSKpg">
        <bpel:from>
          <bpel:literal>
<xform:response>
  <xform:isChainedAfter>
  </xform:isChainedAfter>
  <xform:taskMetaData>
```

182

```xml
<xform:taskId>
</xform:taskId>
<xform:taskState>
</xform:taskState>
<xform:taskType>
</xform:taskType>
<xform:description>
</xform:description>
<xform:processId>
</xform:processId>
<xform:creationDate>
</xform:creationDate>
<xform:userOwner>
</xform:userOwner>
<xform:roleOwner>
</xform:roleOwner>
<xform:claimAction>
  <xform:user>
  </xform:user>
  <xform:role>
  </xform:role>
</xform:claimAction>
<xform:revokeAction>
  <xform:user>
  </xform:user>
  <xform:role>
  </xform:role>
</xform:revokeAction>
<xform:saveAction>
  <xform:user>
  </xform:user>
  <xform:role>
  </xform:role>
</xform:saveAction>
<xform:completeAction>
  <xform:user>
  </xform:user>
  <xform:role>
  </xform:role>
</xform:completeAction>
<xform:formUrl>
```

```
    </xform:formUrl>
    <xform:failureCode>
    </xform:failureCode>
    <xform:failureReason>
    </xform:failureReason>
    <xform:userProcessCompleteSOAPAction>
    </xform:userProcessCompleteSOAPAction>
    <xform:isChainedBefore>
    </xform:isChainedBefore>
    <xform:previousTaskId>
    </xform:previousTaskId>
    <xform:userProcessEndpoint>
    </xform:userProcessEndpoint>
    <xform:userProcessNamespaceURI>
    </xform:userProcessNamespaceURI>
  </xform:taskMetaData>
  <xform:status>OK</xform:status>
  <xform:errorCode>
  </xform:errorCode>
  <xform:errorReason>
  </xform:errorReason>
</xform:response>
          </bpel:literal>
        </bpel:from>
        <bpel:to>$xformNotifyTaskCompletionResponseMsg.root</bpel:to>
      </bpel:copy>
      <bpel:copy          bpmn:label="$sendWeatherEmailSendRequestMsg
out:_r9v_oGzFEdy8_fgUbrSKpg">
        <bpel:from>
          <bpel:literal>
<SendWeather:EmailSend>
  <SendWeather:email>
    <SendWeather:Emailto>
    </SendWeather:Emailto>
    <SendWeather:Emailfrom>
    </SendWeather:Emailfrom>
    <SendWeather:Emailsubject>
    </SendWeather:Emailsubject>
    <SendWeather:Emailbody>
    </SendWeather:Emailbody>
  </SendWeather:email>
```

```
</SendWeather:EmailSend>
        </bpel:literal>
      </bpel:from>
      <bpel:to>$sendWeatherEmailSendRequestMsg.parameters</bpel:to>
    </bpel:copy>
  </bpel:assign>
  <bpel:assign                    bpmn:label="InvokeWeather"
bpmn:id="_hOBjAGzEEdy8_fgUbrSKpg">
    <bpel:copy>

<bpel:from>$thisStartWeatherRequestMsg.body/tns:latitude</bpel:from>

<bpel:to>$sendWeatherWeatherSendRequestMsg.parameters/SendWeather:req
uest/SendWeather:Latitude</bpel:to>
    </bpel:copy>
    <bpel:copy>

<bpel:from>$thisStartWeatherRequestMsg.body/tns:longitude</bpel:from>

<bpel:to>$sendWeatherWeatherSendRequestMsg.parameters/SendWeather:req
uest/SendWeather:Longitude</bpel:to>
    </bpel:copy>
    <bpel:copy>

<bpel:from>$thisStartWeatherRequestMsg.body/tns:startdate</bpel:from>

<bpel:to>$sendWeatherWeatherSendRequestMsg.parameters/SendWeather:req
uest/SendWeather:Startdate</bpel:to>
    </bpel:copy>
    <bpel:copy>

<bpel:from>$thisStartWeatherRequestMsg.body/tns:numdays</bpel:from>

<bpel:to>$sendWeatherWeatherSendRequestMsg.parameters/SendWeather:req
uest/SendWeather:Numdays</bpel:to>
    </bpel:copy>
  </bpel:assign>
  <bpel:invoke                                      partner-
Link="processAndWeatherServiceForPortSendWeatherSoapPlkVar"    port-
Type="SendWeather:SendWeatherSoap" operation="WeatherSend" inputVari-
able="sendWeatherWeatherSendRequestMsg"              outputVari-
```

```
able="sendWeatherWeatherSendResponseMsg"    bpmn:label="InvokeWeather"
bpmn:id="_hOBjAGzEEdy8_fgUbrSKpg" />
    <bpel:assign                              bpmn:label="Create"
bpmn:id="_kLbIQGzEEdy8_fgUbrSKpg">
      <bpel:copy>

<bpel:from>$sendWeatherWeatherSendResponseMsg.parameters/SendWeather:
WeatherSendResult/SendWeather:Dwmlout</bpel:from>

<bpel:to>$xformCreateTaskRequestMsg.root/xform:taskInput/xform:input/
xform:weatherreport</bpel:to>
      </bpel:copy>
    </bpel:assign>
    <bpel:invoke   partnerLink="wFmagic_v0TFUGzDEdy8_fgUbrSKpgPlkVar"
portType="xform:Workflow"      operation="createTask"      inputVari-
able="xformCreateTaskRequestMsg"                         outputVari-
able="xformCreateTaskResponseMsg"             bpmn:label="Create"
bpmn:id="_kLbIQGzEEdy8_fgUbrSKpg" />
    <bpel:receive  partnerLink="wFmagic_v0TFUGzDEdy8_fgUbrSKpgPlkVar"
portType="xform:Process"    operation="notifyTaskCompletion"    vari-
able="xformNotifyTaskCompletionRequestMsg"      bpmn:label="Complete"
bpmn:id="_lKc8UGzEEdy8_fgUbrSKpg" />
    <bpel:reply    partnerLink="wFmagic_v0TFUGzDEdy8_fgUbrSKpgPlkVar"
portType="xform:Process"    operation="notifyTaskCompletion"    vari-
able="xformNotifyTaskCompletionResponseMsg"     bpmn:label="Complete"
bpmn:id="_lKc8UGzEEdy8_fgUbrSKpg" />
    <bpel:if>

<bpel:condition>$xformNotifyTaskCompletionRequestMsg.root/xform:taskO
utput/xform:output/xform:SendEmail</bpel:condition>
      <bpel:sequence>
        <bpel:assign                         bpmn:label="EmailInvoke"
bpmn:id="_rrY5wGzEEdy8_fgUbrSKpg">
          <bpel:copy>

<bpel:from>$xformNotifyTaskCompletionRequestMsg.root/xform:taskOutput
/xform:output/xform:EmailSubject</bpel:from>

<bpel:to>$sendWeatherEmailSendRequestMsg.parameters/SendWeather:email
/SendWeather:Emailsubject</bpel:to>
          </bpel:copy>
```

```
          <bpel:copy>

<bpel:from>$xformNotifyTaskCompletionRequestMsg.root/xform:taskOutput
/xform:output/xform:Emailfrom</bpel:from>

<bpel:to>$sendWeatherEmailSendRequestMsg.parameters/SendWeather:email
/SendWeather:Emailfrom</bpel:to>
          </bpel:copy>
          <bpel:copy>

<bpel:from>$xformNotifyTaskCompletionRequestMsg.root/xform:taskOutput
/xform:output/xform:Email_to</bpel:from>

<bpel:to>$sendWeatherEmailSendRequestMsg.parameters/SendWeather:email
/SendWeather:Emailto</bpel:to>
          </bpel:copy>
          <bpel:copy>

<bpel:from>concat($xformNotifyTaskCompletionRequestMsg.root/xform:tas
kOutput/xform:output/xform:EmailBody,   $sendWeatherWeatherSendRespon-
seMsg.parameters/SendWeather:WeatherSendResult/SendWeather:Dwmlout)</
bpel:from>

<bpel:to>$sendWeatherEmailSendRequestMsg.parameters/SendWeather:email
/SendWeather:Emailbody</bpel:to>
          </bpel:copy>
        </bpel:assign>
        <bpel:invoke                              partner-
Link="emailServiceAndProcessForPortSendEmailSoapPlkVar"        port-
Type="SendWeather:SendEmailSoap"   operation="EmailSend"   inputVari-
able="sendWeatherEmailSendRequestMsg"                    outputVari-
able="sendWeatherEmailSendResponseMsg"      bpmn:label="EmailInvoke"
bpmn:id="_rrY5wGzEEdy8_fgUbrSKpg" />
          <bpel:empty                    bpmn:label="EventEndEmpty"
bpmn:id="_SylQUGzIEdy8_fgUbrSKpg" />
      </bpel:sequence>
      <bpel:else>
        <bpel:sequence>
          <bpel:empty                  bpmn:label="EventEndEmpty"
bpmn:id="_SIjb4GzIEdy8_fgUbrSKpg" />
        </bpel:sequence>
```

187

```
  </bpel:else>
 </bpel:if>
 </bpel:sequence>
</bpel:process>
```

3. *Screen shot of the User Form of the Implemented Process of the WCS With Approval Task – Input for the Weather Service*

Fig F.3 shows the user form to input the weather parameters for retrieving the weather report.

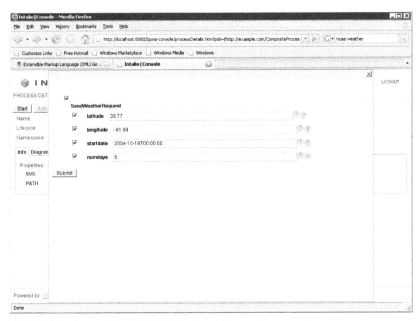

Fig. F.3. User form of the of the WCS to input the weather parameters..

188

4. *Screen shot of the Administration Page of the Implemented Process of the WCS With Approval Task – Shows that a Task is on the Employee Queue*

Fig F.4 shows the administration page of the approval task. It shows that a task is in the employee queue.

Fig. F.4. Administration page of the WCS.

189

5. *Screen shot of the Administration Page of the Implemented Process of the WCS With Approval Task – Approval Form*

Fig F.5 shows the administration page with the approval form.

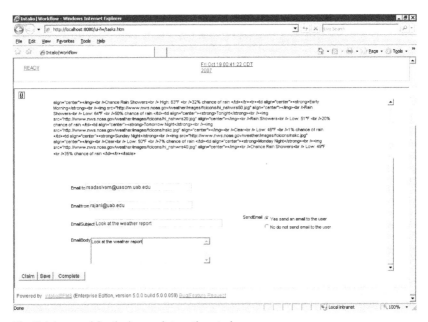

Fig. F.4 Approval for the human interactions task.

6. *Screen shot of the Administration Page of the Implemented Process of the WCS With Approval Task – Email Received by User*

Fig F.5 shows the email received by the user.

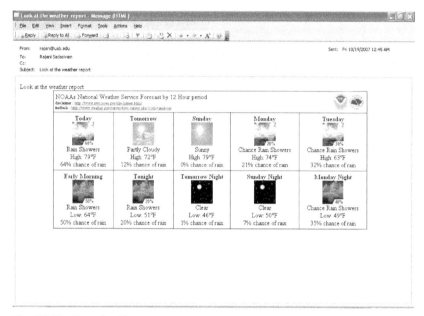

Fig. F.5. Email received by user.

APPENDIX E

SEMANTICS FOR THE GLIP

I provide the KS, RUS, ROS, UPS, INS, and COS for the GLIP case study. KS answers specific needs of service-agents to perform a task. ROS specifies the conditions necessary to carry out a task. ROS describes roles that provide a representation of the set of users that are responsible for a task. UPS provides a representation of the users who are participating in a process and their preferences. INS describes the types of resources that the user requires to complete a task. COS represents the input and output messages of the different tasks in a process. Several approaches can be used to specify the six types of semantics, as discussed in Chapter V. There are five tasks in the GLIP case study: MapViewer, OMIMCheck, Articles, KeywordSearch, and GeneSelection.

1. MapViewer Task

KS for the task is as follows:

1) What are the maps used?

- Maps used are Marshfield Map, the Genes Sequence Map, and the Ideogram Map.
- Gene Sequence Map lists all the genes on the genomic region specified.
- Marshfield map is required to view the chromosome region in the centi-Morgan units (cM).
- Ideogram map is a cytogenetic map and represents the ideogram of the G-banding pattern at the 850-band resolution.

More   information   about   the   maps   is   listed   at
http://www.ncbi.nlm.nih.gov/mapview/static/humansearch.html.

2) What is the format of the genes list?

When Gene Sequence is selected as the Master map, the verbose display (detailed labeling, shown by default) includes arrows to the right of each gene name to indicate its direction of transcription as well as links to the following:

- OMIM - Online Mendelian Inheritance in Man.
- sv - sequence viewer.
- pr – protein.
- dl - view/download sequence data from a chromosome region.
- ev - evidence viewer.
- mm - Model Maker.
- hm   –   HomoloGene.

Additional   information   is   provided   at
http://www.ncbi.nlm.nih.gov/mapview/static/humansearch.html#genes

RUS for the task is as follows:

- No RUS for the task.

ROS for the task is as follows:

- The post-doctoral fellows of the Department of Epidemiology are assigned to the task.

UPS for the task is as follows:

- Dr. Laura Vaughan, a post-doctoral fellow of the Department of Epidemiology, is a user of the task. She uses the NCBI Map View tool to perform the task.

INS for the task is as follows:

- The INS for the task is the NCBI Map View tool.

COS for the task is as follows:

- The input message, which is the map and region required to invoke the Map Viewer tool.
- The   output   message,   which   is   the   genes   list.

2. *OMIMCheck task*

KS for the task is as follows:

1) What is the input genes list?

- See answer 2, format of the genes list in the KS of Map Viewer task.

2) What is the output genes list?

- The output genes list is the list created after the genes without OMIM information are removed.

RUS for the task is as follows:

1) What is the condition to identify the genes with OMIM information?

- In the input genes list, the "LINKS" column indicates the presence or absence of OMIM information. Textually, it is represented as "OMIM" in the "LINKS" column to indicate the presence of OMIM information. An empty field in the "LINKS" column indicates the absence of OMIM information.

ROS for the task is as follows:

- The post-doctoral fellows of the Department of Epidemiology are assigned to the task.

UPS for the task is as follows:

- Dr. Laura Vaughan, a post-doctoral fellow of the Department of Epidemiology, is a user of the task. She uses the Microsoft Excel tool to perform the task.

INS for the task is as follows:

- The Microsoft Excel tool, which is used to view and manipulate the genes list.

COS for the task is as follows:

- The input message, which is the genes list.
- The output message, which is the modified genes list created after the genes without OMIM information are removed.

3. *Articles Task*

KS for the task is as follows:

- No KS for the task.

RUS for the task is as follows:

1) When should I invoke the NCBI OMIM and Pubmed Database?

- Run retrieval scripts on weekends or between 9:00 p.m. and 5:00 a.m. Eastern Time weekdays for any series of more than 100 requests.
- Send e-utilities requests to http://eutils.ncbi.nlm.nih.gov, not the standard NCBI Web address.
- Make no more than one request every 3 seconds.
- Use the URL parameter email, and the tool used for distributed software, so that NCBI can track your project and contact you if there is a problem.
- NCBI's Disclaimer and Copyright notice must be evident to users of your service. National Library of Medicine (NLM) does not claim the copyright on the abstracts in PubMed; however, journal publishers or authors may. NLM provides no legal advice concerning distribution of copyrighted materials, so consult your legal counsel.

ROS for the task is as follows:

- The post-doctoral fellows of the Department of Epidemiology are assigned to the task.

UPS for the task is as follows:

- Dr. Laura Vaughan, a post-doctoral fellow of the Department of Epidemiology, is a user of the task. She uses the file upload form tool to upload the genes list.

INS for the task is as follows:

- The OMIM and PubMed database service that is used to query for the OMIM and Pubmed articles.
- The file upload form that is used to upload the genes list for querying the OMIM and PubMed database service.

COS for the task is as follows:

- The input message, which is the genes list.
- The output message, which is the articles retrieved from the OMIM and PubMed database service.

4. *KeywordSearch task*

KS for the keyword search task is as follows:

1) What is the input genes list?
   - See answer 2 KS output genes list in the OMIM check task.
2) What is the format of the articles?
   - The format of the articles is the gene information, such as gene id and gene name, and the associated articles retrieved from the OMIM and Pubmed information.
3) What are the relevant search keywords?
   - The user role history provides some example of search keywords.
4) What are the search results?
   - The search results contain the results of the keyword of interest search on the articles.

RUS for the task is as follows:

- No RUS for the task.

ROS for the task is as follows:

- The post-doctoral fellows of the Department of Epidemiology are assigned to the task.

UPS for the task is as follows:

- Dr. Laura Vaughan, a post-doctoral fellow of the Department of Epidemiology, is a user of the task. She uses the Microsoft Word tool to perform the task.

INS for the task is as follows:

- Microsoft Word tool that is used to view the articles and store the results of the keyword of interest search.

COS for the task is as follows:

- The keywords of interest.
- The search results for the keyword of interest search.

5. *GeneSelection task*

KS for the genes search task is as follows:

1) What is the input genes list?

- See answer 2 KS output genes list in the OMIM check task.

2) What is the format of the articles?

- See answer 3 KS format of articles in the articles task.

3) What is the output genes list?

- The output genes list is a list of the genes whose PubMed and OMIM summaries do not match the keyword of interest relevant to the study.

4) What are the search results?

- See KS answer 5 for the search results of the keyword search task.

RUS for the keyword search task is as follows:

1) What is the condition to select genes for continued study?

- Genes whose PubMed and OMIM summary matched the keyword of interest relevant to the study are the set of genes for continued study.

ROS for the task is as follows:

- The post-doctoral fellows of the Department of Epidemiology are assigned to the task.

UPS for the task is as follows:

- Dr. Laura Vaughan, a post-doctoral fellow of the Department of Epidemiology, is a user of the task. She uses the Microsoft Word tool to perform the task.

INS for the task is as follows:

- Microsoft Word tool is the tool that is used to view the articles and store the results of the keyword of interest search.

COS for the task is as follows:

- The input genes list.
- The matched articles, which is the results of the keyword search task.
- The output genes list.

VDM

Verlag
Dr. Müller

Wissenschaftlicher Buchverlag bietet

kostenfreie

Publikation

von

wissenschaftlichen Arbeiten

Diplomarbeiten, Magisterarbeiten, Master und Bachelor Theses
sowie Dissertationen, Habilitationen und wissenschaftliche Monographien

Sie verfügen über eine wissenschaftliche Abschlußarbeit zu aktuellen oder zeitlosen
Fragestellungen, die hohen inhaltlichen und formalen Ansprüchen genügt,
und haben **Interesse an einer honorarvergüteten Publikation**?

Dann senden Sie bitte erste Informationen über Ihre Arbeit per Email
an info@vdm-verlag.de. Unser Außenlektorat meldet sich umgehend bei Ihnen.

VDM Verlag Dr. Müller Aktiengesellschaft & Co. KG
Dudweiler Landstraße 125a
D - 66123 Saarbrücken

www.vdm-verlag.de

www.ingramcontent.com/pod-product-compliance
Lightning Source LLC
LaVergne TN
LVHW022310060326
832902LV00020B/3373